Detlef Egbert Ricken · Wolfgang Gessner (Eds.)

Advanced Microsystems for Automotive Applications 99

Springer
Berlin
Heidelberg
New York
Barcelona
Hong Kong
London
Milano
Paris
Singapore
Tokyo

Detlef E. Ricken · Wolfgang Gessner (Eds.)

Advanced Microsystems for Automotive Applications 99

With 250 Figures

Dr. Detlef Egbert Ricken
VDI/VDE-Technologiezentrum Informationstechnik GmbH
Rheinstr. 10B
D-14513 Teltow
e-mail: ricken@vdivde-it.de

Wolfgang Gessner
VDI/VDE-Technologiezentrum Informationstechnik GmbH
Rheinstr. 10B
D-14513 Teltow
e-mail: gessner@vdivde-it.de

ISBN 3-540-65183-7 Springer-Verlag Berlin Heidelberg New York

Library of Congress Cataloging-in-Publication Data

Advanced Microsystems for Automotive Applications 99 / Eds.: Detlef Egbert Ricken; Wolfgang Gessner. – Berlin; Heidelberg; New York; Barcelona; Budapest; Hongkong; London; Milano; Paris; Singapore; Tokyo: Springer, 1999
ISBN 3-540-65183-7

Typesetting: Camera-ready by authors
Cover-Design: de'blik, Berlin

SPIN: 10698619 68/3020 - 5 4 3 2 1 0 - Printed on acid-free paper

Preface

Modern societies are characterised by an increase of mobility. Therefore, the requirements of present-day transport systems become more and more complex and new technology applications in the vehicle, the infrastructures and in logistics will accelerate developments. Comfort, security, engine performance and emissions as well as design play a significant role for developers and user. Especially modern automobiles have to correspond to these demands. Microsystems are indispensable for these features.

Microsystems or MEMS - the combination of (mechanical) sensing elements and electronic data processing into a single system - are key elements in the „intelligent car“. The increasing interest for microsystem applications in the automotive sector shows that this topic deserves to be discussed on a broad international level and emphasises the role of Microsystems for the development of a global economy. Microsystems are the invisible assistants who help to reduce engine emissions and to improve performance, who assist the driver to keep distance to the vehicles in front and enable him to get more easily and safely through the traffic. The contributions of this publication reflect the discussion during the 3rd Conference on Advanced Microsystems for Automotive Applications carried out in Berlin on March 18/19, 1999. They show new technological solutions in areas like traffic management, communication, engine management and safety.

We would like to express our sincere thanks to all authors who contributed to this discussion with their papers and to the members of the Honorary Committee and Steering Committee for their engagement and support. We also would like to thank the Commission of the European Community for their financial support. Special thanks are addressed to Mr. Roger Grace from Roger Grace Associates for his marketing support in the US, to Mrs. Schillings and Mrs. Thiel, who took care of bringing this book to publication.

Teltow, January 1999

Detlef Egbert Ricken　　　　Wolfgang Gessner

Table of Contents

Microsystems for Automotive Applications 1
W. Gessner et.al., VDI/VDE-Technologiezentrum Informationstechnik GmbH

Session 1: Microsystems for Traffic Management and Communication

Intelligent Car Communication Systems 13
D. Tu, Texas Instruments

The Application of PC Technology in Automobiles 25
J. Casazza, Intel Corporation

Advanced Microphones for Handsfree Communication and Voice Recognition 43
R. Frodl, Ruf-Electronics GmbH

Using In-Vehicle Systems and 5,8 GHz DSRC to Improve Driver Safety and Traffic Management 63
O. Clair, Renault

Session 2: Systems and Technologies for Media Control and Engine Management

Intelligent Engine Control & Diagnostic Using the Ionization Current Sensing Technology 71
J. Duhr, Delphi Automotive Systems Luxembourg S. A.

Micro-Injection System Using Ultrasonic Vibrations for Drop on Demand Ejection 87
L. Lévin et.al., Technocentre Renault

Integrated Engine-CVT Control 97
M. Nozaki et.al., Nissan Motor Co. Ltd.

Production of Light Wave Connectors, Hermetic Housings, Heat Sinks etc. by the MIM Process 109
R. Hardt, Industriekontor Rolf Hardt

Advanced Packaging and Interconnection Technologies for Automotive Applications 111
P. Sommerfeld et.al., Philips GmbH

A Low Cost, Fully Signal Conditioned Pressure Sensor Microsystem with Excellent Media Compatibility 121
R. Grelland et.al., SensoNor asa

Session 3: Microsystems Related to Safety Aspects

Occupant Classification System for Smart Restraint Systems 133
K. Billen et.al., I.E.E. International Electronics & Engineering

Concept for Intelligent Rear Light Aiming for Compensation of Environmental Effects 147
W. Robel et.al., Reitter & Schefenacker GmbH & Co. KG

Miniaturized Scanning Laser Radar for Automotive Applications 173
M. Monti et.al., CSEM SA

100.000 Pixel 120dB Imager for Automotive Vision 183
T. Lulé et.al., Silicon Vision GmbH

Development of RF Technology for Automotive Radar and Mobile Broadband Communication 195
F. Rehme et.al., DaimlerChrysler Aerospace AG

Session 4: Micromachined Sensors and Actuators

A Low Cost Micro-Inertial and Flow Sensors Based on the Direct Integration Technology 199
D. Haronian, Tel-Aviv University

Novel Rotation Speed Measurement Concept for ABS Appropriated for Microsystem Creation 215
S. Yurish et.al., Institute of Computer Technologies

A Novel Technology Platform for Versatile Micromachined Accelerometers 225
S. Toelg et.al., EG&G Heimann Optoelectronics GmbH

A Low Cost Angular Rate Sensor for Automotive Applications in Surface Micromachining Technology 239
R. Schellin et.al., Robert Bosch GmbH

A New Generation of Micromachined Accelerometers for Airbag Applications 251
M. Aikele et.al., TEMIC Sensorsysteme

Integrated Surface Micromachined Gyro and Accelerometers for Automotive Sensor Applications 261
B. Sulouff, Analog Devices Inc.

A New Microelements Electrostatic Actuator for Automotive Applications 271
M. Pizzi et.al., Fiat Research Center

Poster presentations

The Application and Evaluation of a Novel Engine Management System Based on Intelligent Control and Diagnostics Algorithms and Utilising Innovative Sensor Technology 279
A. Truscott et.al., Ricardo Consulting Engineers Ltd.

System for Radio-Controlled Car Clocks 285
R. Polonio et.al., TEMIC Semiconductors

Fuel Injection Engine Diagnosis 289
P. Ripoll et.al., LAMII/CESALP

Monolithic Pressure Sensor System with Digital Signal Processing 297
J. Schuster, Motorola

An MCM Microcontroller for Automotive Applications 309
A. Simsek, Fraunhofer Institut FhG-IZM

List of Contact Addresses 313

AMAA Honorary Committee

AMAA Steering Committee

Microsystems for Automotive Applications

W. Gessner, A. Hilbert, Th. Köhler, D.E. Ricken
VDI/VDE-Technologiezentrum Informationstechnik GmbH,
Teltow

Abstract

We briefly describe the history of some major automotive developments. Electric systems were already part of the first „automobile". While ignition systems were the first electric systems, radios built with vacuum tubes were the first electronic systems. The subsequent use of semiconductors enabled a wide variety of intelligent functions. Today complex Microsystems, the combination of e.g. electronical and mechanical functions are a key component of all vehicles.

Introduction

The birth of combustion engines took place without any electrical systems. However the „Benz Patent-Motorwagen" was already equipped with a 4 stroke combustion-engine, in which ignition was triggered by an electric ignition system. Details of the engine can be found in various historical monographs [1]. Electric ignition systems were developed by engineers of the Eyquem-Factory in 1895. In 1902 Robert Bosch invented the spark plug. For many decades the ignition system was a very simple electro-mechanical system.
In the 1930ies the AM and later the FM radio was the first electronic component that was built into cars. These early entertainment systems were built with conventional electronic elements: simple tuneable capacitors for the receiver and vacuum tubes for the amplifier [2].

The investigation of semiconducting materials started in the early nineteenth century. Researchers investigated the different behaviours of metals vs. other materials like silicon (Si), germanium (Ge) or compound materials In contrast to metals these materials show an increasing conductivity with increasing temperature. The *energy-band-model*, a theoretical model for the explanation of these phenomena was developed by Brillouin. Based on the theoretical model Bardeen and Brattain realised a first transistor (contraction of *transfer (of an electrical signal across a) resistors*) in 1947 [3].

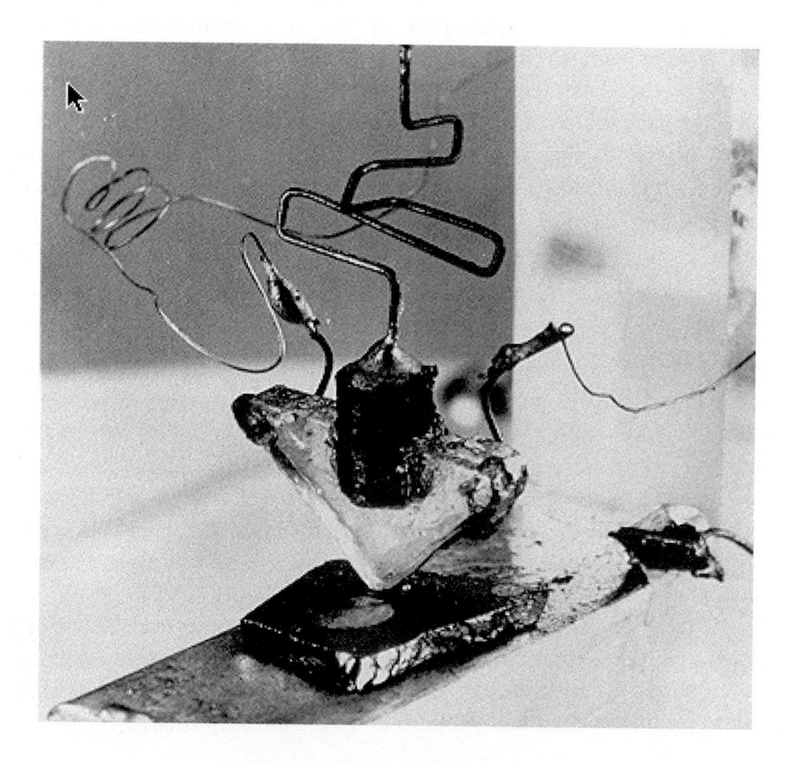

Fig. 1: The first transistor [3]

The theoretical description of the related interface (pnp-junctions) was made by Shockley [4]. Compared to vacuum tubes, transistors offered tremendous advantages. They were much smaller and therefore faster and they need no heated electrodes. Thus energy consumption was much smaller. Between 1948 and 1970 semiconductor electronics with discrete transistors and diodes started to replace the older technology.

The first monolithic integrated circuit (IC) was designed by Jack Kilby [5] and manufactured by Texas Instruments, Inc. in 1959. During the 70'ies the IC-technology emerged. The Intel 8080, a milestone in Intels processor history, became the brains of the first personal computer--the Altair-- a microcomputer with 6.000 transistors and 2 MHz clock speed, 25 years later Intel's 7.5 million-transistor Pentium® II processor or Motorola's K6-family were available. Parallel to the development of more and more powerful processors in particular Japanese companies focused on the development of memory chips. In the late 80ies the 1MB Chip became available. Today 256MB chips are going into mass production.
With the availability of monocrystalline Silicon and a well established process technology the idea emerged to use Silicon for electronic as well as for mechanic devices. Based on the anisotropic etching technology the first micromechanic devices were discussed. [6]. The first demonstrators, pressure sensors and accelerometers were produced within only a few years. A decade later a variety of micromachined sensors and passive elements were available as industrial mass products. Today the MEMS (:= micro electro mechanical systems), MST (:= microsystem technology) or Micromachining are common names for a whole family of products and technologies. Each name has a different emphasis but all names describe intelligent components and systems made with common semiconductor technologies plus additional process steps.

Ignition: An Electrical System enables Combustion Engines

Drivers expect reliable cars, offering mobility. Thus automotive applications require mature and reliable technologies. These technologies were developed by mechanical-engineers and innovators from all over the world. In the beginning engineers from Europe and the United States used their excellent knowledge of steam engines and thermodynamic principles for the construction of their engines.

The experiments with various engine types revealed that combustion engines were the most adequate solution for vehicles. Therefore electrical components became necessary. A combustion engine requires a reliable ignition system. Carl Friedrich Benz, the inventor of the Benz Motorwagen stated that the ignition was the "problem of problems". In his autobiography he wrote that the best engine design is useless if the spark is missing ("Wenn der Funke ausbleibt, sind die besten Motorkonstruktionen keinen Pfifferling wert" [7]). Therefore his "Benz-Patent-Motorwagen" used a Lenoir-High-Voltage-Ignition with a Ruhmkorff-Spark-Inductor. The required energy was provided by a galvanic element filled with chromium acid. The output voltage of the battery was 4 Volts.
This state-of-the-art ignition system enabled the operator to drive approximately 10 km before the battery needed to be replaced. The reliability of the system, equipped with a Wagner-Hammer as crucial element need not be discussed.
Given these problems it is evident that a tremendous amount of work was

Fig. 2: The Benz Patent-Motorwagen [8]

neccssary in order to develop the toy of early technical enthusiasts into a reliable tool for everyday life. This objective was stated very early e.g. by the company Benz. In 1888 they had the slogan "*Steer, stop and brake more easily and safely as in conventional vehicles - no special operation required.*"

The ignition system used by Benz offered clear advantages compared to all other possibilities to trigger the combustion process. The only competitive system was favoured and optimised by Gottlieb Daimler. He used a hot tube, heated by burning gasoline, for the ignition of the air/fuel mixture inside the cylinder. The advantage of this ignition system were reliability and simplicity but after 1893 new requirements could not be satisfied. The incandescent plug could not be used for fast running engines. Therefore the idea of Carl Friedrich Benz proved to be right, even if the contemporary author Baudry de Saunier [9] wrote that "electricity and despair are the same".

Subsequent development was focused on improving the electric ignition. The breakthrough came in 1887 when the magnetic low voltage ignition system was developed by Robert Bosch. This invention finally merged mechanical and electrical systems in the automobile. Furthermore it was the birth of a completely new industry: The electrical supplier industry for the automobile manufacturers.

The construction of a breaker, the invention of "modern" spark plugs by Eyquem (1899) and Bosch (1902) without moveable parts inside the cylinder and the use of high-voltage systems were the next improvements to the ignition system.

In particular in the US the automobile was used by an increasing number of customers. For example more than a million Ford Model T's were sold until 1915 [10]. Many drivers were technically inexperienced. They demanded reliable and easy to use systems. Therefore also electrical starters were necessary. This required powerful accumulators which soon became available. Together with these batteries, automobiles could be equipped with simple ignition systems consisting of breaker (de Dion-Bouton, 1895), capacitor (Fizeau, 1853) distributor (Lenoir 1860) and ignition-coil (1925) allowing safe operation even at low engine speed. After 1935 such systems were used world-wide.

The link between the electric system and a rechargeable energy buffer was an elegant and technically obvious solution if the buffer could be recharged during the drive. This construction requires knowledge about electrical engines. Carl Friedrich Benz tried in 1882 to transfer generators designed for stationary operation into the automobile. However it took until 1900 to develop generators suitable for automotive applications. Problems with the mechanical design were solved within a short time and in 1904 the first suitable generators were provided by C. A. Vanderrell Co. (CAV), a company in the UK. But the problem of a regulator for the generator was unsolved. This was even more important since in 1915 Ford introduced electric lighting into the automobile.

The voltage regulation with a contact regulator developed by Tirril for stationary generators had significant advantages. After some modifications by Robert Bosch it became the leading design for the next 15 years.

The evolution of vehicles was accompanied by an increasing number of electric systems. The growing demand for electric energy could not be satisfied with larger or faster running DC generators. Therefore AC generators were needed. They used an external Selenium rectifier to provide DC voltage.
The Selenium rectifier was the first element using semiconducting materials in the vehicle. The link between automotive and semiconductor technology was established. The next step was made in the early 60ies when three phase current generators were introduced into the automobile. These generators use Silicon diodes for rectification and the use of semiconductor devices became common technology. With semiconductor devices electronic systems could find their way into the car. Earlier attempts to use electronics in the automobile had failed. For example a patent dated 1920 proposed the usage of vacuum electronics in the ignition system. But vacuum tubes were not suitable for most automotive applications and therefore this and many other brilliant ideas could not be realised. There was only one system were vacuum electronics were used in the automotive environment: The radio.

Information and Entertainment: The first Electronic System

The Idealwerke which later became the Blaupunkt Werke developed the first European car-radio in 1932. The unit was able to receive AM stations while the vehicle was driving. Since the receiver was too large to be close to the driver it was operated via remote controls attached to the steering column.

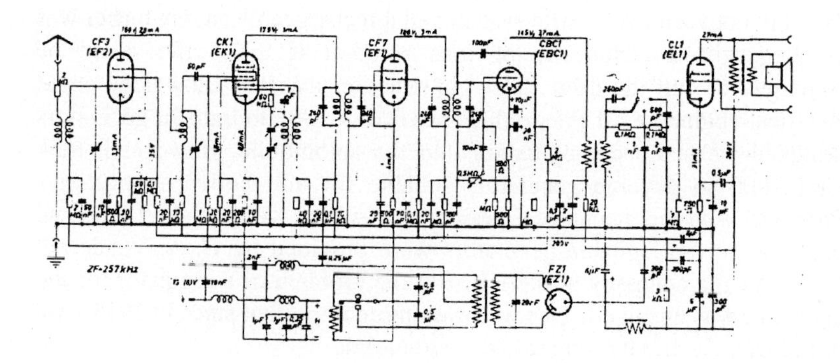

Fig. 3: The schematic diagram of AS 5 A 76, one of the first car-radios

While the vacuum tube was not compatible with a variety of automotive applications the transistor was. In the beginning the performance of the transistor was very poor but the size and its low operation temperatures were an extreme advantage. Thus as early as 1953 a first patent proposing the use of a transistor controlled ignition coil was filed [11]. Subsequent patents included engine speed and load sensitive regulation of the ignition system and the first transistor ignition system was introduced in the early 60ies.
The first power transistors enabled engineers to improve the design of the electric network between generator and battery. Better recharge characteristics and voltage stability were achieved at the end of the 60ies since the available transistors allowed the realisation of stable regulators. Ford, GM, Lucas and Siemens presented a number of optimised concepts and brought these concepts onto the market.
In Europe flashing warning lights became mandatory in the late 60ies. This was the trigger to use transistor electronics for the operation of flashing lights. Flashing lights were introduced by Ford in 1916 but until then they were triggered by electromechanical relays. These systems disappeared very quickly after the introduction of transistor circuits.
The requirements for automotive electronics are very high. Therefore engineers started to discuss the potential of integrated circuits, introduced in 1958 by J. S. Kilby from Texas Instruments. The first system that used these new IC's was the radio. Since approx. 1967 car-radios have been equipped with IC's. Other systems followed within only a few years e.g. regulator IC's (1970) flash-light triggers (1969) IC's for fuel injection and dashboard-instruments. It is interesting to mention that already in these early years the first engineers proposed to use electrical signals from these systems to supply safety systems like ABS-systems with relevant data.
The next phase started with the development of microprocessors. These devices offered the possibility to register complex processes and to process the data in a flexible way. This enabled the development of a complete electronic system for engine management with load-dependent fuel injection and ignition. The first engine management system equipped with a microprocessor was presented in 1979.
Electronic systems became omnipresent in the vehicle while scientists realised that semiconducting materials have excellent mechanical properties too....

Smart Systems for the Vehicle

Information technology was the driving force of the semiconductor industry. However with the availability of electronics automotive

engineers started to use smart electronic systems to integrate new features into the automobile.
Electronic ignition systems became available. The evolution of these systems lead to complex engine management systems. Equipped with various sensors a modern electronic control unit (ECU) enables the realisation of powerful and economic engines with fuel injection, emitting only a fraction of poisonous exhaust gases compared to old carburettor engines.
To avoid accidents it is important to enable the driver to control the vehicle in critical situations. Sophisticated mechanical concepts for frames and axles guarantee a high safety level for today's vehicles. While the advanced mechanical structure is the basis for safe driving, electronic systems today can recognise dangerous situations even before the driver. Equipped with smart sensors, intelligent signal processing and fast actuators they can assist the driver to keep the vehicle under control and to stop it as soon as possible. Antilock brake systems (ABS) were the first of these products. Today they are part of a complete vehicle dynamics control system.
The sensors for ABS and more sophisticated systems for vehicle dynamic control systems use information about wheel speed, measured with magnetic sensors. A wide variety of earlier sensors used the Hall-Effect. In a changing magnetic field this effect generates a voltage. The disadvantage of these kinds of sensors is the sensitivity to external fields. More recent systems use magnetoresistive sensors. A new alternative using an „active microsystems" is discussed in the paper by Yurish et. al [12].
Inertial sensors are also required. Micromachined accelerometers are mass produced by a variety of companies. Compared to „simple" accelerometers micromachined gyroscopes and yaw-rate-sensors are very complex devices. The rotation is measured by the Coriolis force on a rotating or oscillating system. Therefore the gyroscope needs an actuator to stimulate the oscillation. Since the electric output signal is much smaller than the input for the actuator these devices require very sophisticated driver electronics and amplifier technology. Both accelerometers and gyros play an important role for many automotive systems. Therefore new developments will improve these devices. (Surface) micromachining is the technology used for many designs [13], [14], [15]. Today capacitive signal readout is almost standard but e.g. resonant detection principles have yet to be evaluated [13].
In the event of an accident the construction of the car body is the outstanding safety feature of a vehicle. Crunch zones and other energy absorbing features of the car body cannot avoid injuries due to impacts of the passenger into parts of interior live cell. Simple seat belt systems are the first and most important element of a sophisticated protection chain.

Today pretensioners and airbags are a widely used feature for passenger protection. The availability of mass produced silicon micromachined sensors enabled the introduction of these safety features even into low priced cars. Accelerometers and gyros have been discussed already. However the situation appropriate deployment of airbags requires more. It requires knowledge about the accident even before it happens: Thus pre-crash-sensors can activate the various protection systems and the response of the total system can be tailored to the accident. Radar sensors and other systems for obstacle detection can be used for this purpose.
Another important parameter is the situation inside the car. Which seats are occupied and what are the passengers doing? Optical, ultrasonic, Radar and other technologies are under evaluation. All these technologies have their specific advantages and problems. It might be necessary to combine these detection principles with an array of pressure sensors in the seat. The potential of pressure sensor arrays is enormous [16] and a sensor mat can provide important information for an intelligent passenger protection system.
Customers spend a significant portion of their time in the vehicle. Thus driving should not only be safe. It should be convenient as well. Smart systems can do a lot for convenient driving. Simple heating systems evolved to advanced climate control systems. The simple radio has become an audio-entertainment-system and will develop into a mobile multi media system for navigation, general information and entertaining. Thus the radio, the first electronic component of the automobile is transferred into a centre of new interactive services and outstanding features. This requires new technologies: If the driver is to use these services while he is driving new interfaces have to be developed. Microphones for external communication and for voice recognition systems are required. Microsystem technologies enable designers to develop advanced in-car-microphones e.g. for handsfree communication [17].
Traffic reports are very important information. Today they are received via radio. Some mobile phone providers offer additional services. A complex intelligent transportation system needs more. Information has to be transferred from the vehicle to the intelligent highway and back. Service providers need detailed information of current road conditions in order to recommend the best routing for their customers. Various aspects of the required technologies will be discussed in this book.

Up-and-coming Technologies

Like microelectronics, microsystems or MEMS make a lot of intelligent functions possible. Without them modern cars are unimaginable. And the importance of microsystems will increase. A questionnaire, mailed to a number of experts from automobile manufacturers and suppliers revealed that the importance of microsystems will grow even faster than the importance of „pure“ microelectronics. It seems to be realistic that for automotive applications the added value of microsystems will grow to more than 5% within the next 10 years. The main drawback for the introduction of further microsystems is the cost. Required base technologies seem to be available waiting to be used in new products. Functional categories of these products and some recent developments are discussed in a review article by Jost [18] Fig. 4 compares the potential of some MEMS-technologies and devices.

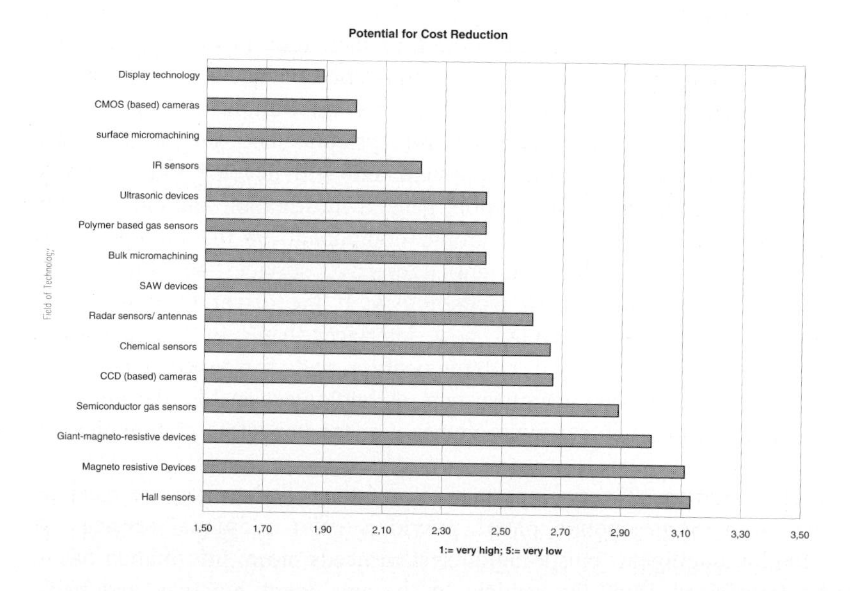

Fig. 4 Estimated Potential for Cost Reduction; Results of a survey

Displays seem to be the most important microsystems for new generations of automobile. This is not surprising since new displays can reduce costs and enable a lot of new functions (e.g. head up displays for automotive applications)

A very interesting result of the survey is the high potential of polymer based gas detectors. Although these devices today are only suitable for limited applications the estimated potential is very high. This short review demonstrates that microelectronics enabled new revolutionary features. Microsystems will change the automobile even more dramatic. But this revolution is only at its beginning.

Acknowledgement

The authors like to thank Mr. Hans Straßl, Kurator Abt. Kraftfahrzeuge Deutsches Museum München for his information about historical vehicles.

Literature

[1] F. Sass, Geschichte des deutschen Verbrennungsmotorenbaues von 1860 bis 1918, Springer-Verlag, Berlin 1962; H. de Boer, Th. Dobbelaar, G. Mom, Das Auto und seine Elektrik, Motorbuch Verlag, 1990; G. Conzelmann, U. Kincke "Mikroelektronik im Kraftfahrzeug", Springer-Verlag, Berlin 1995

[2] http://www.blaupunkt.de/his/his.htm

[3] J. Bardeen and W.H. Brattain, The transistor, a semiconductor triode, Phys. Rev., **74**, 230 (1948)

[4] W. Shockley, The theory of p-n junction in semiconductors and p-n junction transistor, Bell Syst. Tech. J. **28**, 435 (1949)

[5] US-Patent # 3,138,743

[6] K.E. Petersen, Micromechanical membrane switches on Silicon, IBM J. Res. Develop. **23,** 376 (1979); K.E. Petersen, Silicon as mechanical material, Proc IEEE **70**, 420 (1982).

[7] C. Benz, - Lebensfahrt eines deutschen Erfinders. Erinnerungen eines Achzigjährigen. Koehler & Amelang, Leipzig, 1925

[8] http://www.detnews.com/1997/autos/9708/18/08170022.htm

[9] B. de Saunier: Grundbegriffe des Automobilismus. Wien, Leipzig 1902

[10] http://www.modelt.org/tcars.html

[11] US-Patent 2,528,589 Ignition system for an internal combustion engine, 8.1.1953

[12] S.E. Yurish: Novel Rotation Speed Measurement Concept for ABS Appropriated for Microsystem Creation in: Advanced Microsystems for Automotive Applications 99, Springer Verlag 1999

[13] M. Aikele: A New Generation of Micromachined Accelerometers for Airbag Applications in: Advanced Microsystems for Automotive Applications 99, Springer Verlag 1999

[14] B. Sulouff: Integrated Surface Micromachined Gyro and Accelerometers for Automotive Sensor Application in: Advanced Microsystems for Automotive Applications 99, Springer Verlag 1999

[15] R. Schellin: A Low Cost Angular Rate Sensor for Automotive Applications in Surface Micromachining Technology in: Advanced Microsystems for Automotive Applications 99, Springer Verlag 1999

[16] K. Billen: Occupant Classification System for Smart Restraint Systems in: Advanced Microsystems for Automotive Applications 99, Springer Verlag 1999

[17] R. Frodl: Advanced Microphones for Handsfree Communication and Voice Recognition in: Advanced Microsystems for Automotive Applications 99, Springer Verlag 1999

[18] K. Jost, SAE and intelligent vehicles, Automot. Eng. Int., IV3, October 1998

Intelligent Car Communication Systems

Dung Tu
Texas Instruments Deutschland, Haggertystr. 1, 85356 Freising, dung-tu@ti.com

Abstract: Car telephone, car entertainment, navigation system and board computers are optional accessories of a car that are available today. However, these parts operate independently and do not support each other. If the car-phone can talk to the navigation system and there is a service provider who monitors the position of the car to inform about traffic congestion then the system will be smart enough to recommend a new route. Supposed that the bandwidth of the cellular phone is sufficient, the system can download an up-to-date map from a server instead of using a map from a CD-ROM. Applications like emergency calls, breakdown assistance, email, Internet services, remote diagnostics, etc. are also useful. The paper discusses various aspects of an intelligent car communication system that are relevant to achieve optimal performance and cost.

Keywords: Communication, Entertainment, GSM – based Services, GATS, WAP

1. Introduction

In our information society, mobile phones, computers, Internet, TV and video are penetrating nearly all sectors of daily life. The question is how these applications will find inroad into the car. Telematics, Intelligent Transport Systems, Infotronics are the headlines. Because consumers are only willing to pay for a product or service if they perceive it as desirable, it is important not only to consider the technical feasibility but also to look at the big picture. The success of new systems depends very much on affordable prices and useful services. This requires a high degree of integration and a common sense for standardization so that the various components of systems and services fit to each other.

2. Value chain

Figure 1 shows the value chain how the content of a service makes use of an infrastructure and an on-board terminal to reach a user to satisfy his needs.

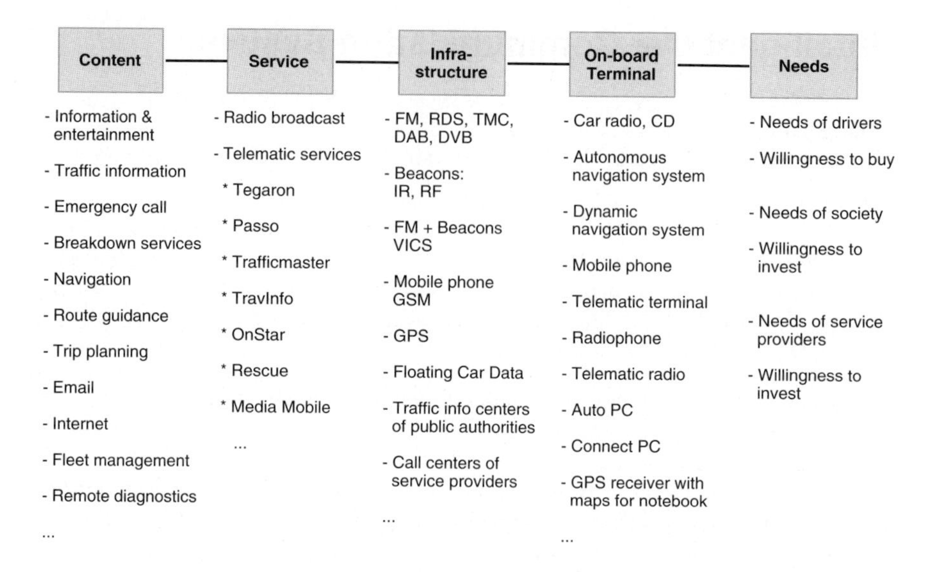

Fig.1: The value chain

2.1 Needs Determine Service Content

Although at the end of the chain, the starting point should be the needs of drivers and their willingness to pay for useful services and equipment. The needs determine the content of services. Entertainment and information, including traffic information, count as basic needs. Emergency call, breakdown assistance, navigation and route guidance, especially in case of traffic congestion are certainly desirable. Personal communication and information like email, trip planning, etc. are useful if they are implemented in a sensible way. Remote diagnostics might be useful for carmakers to offer new services. There is also a need from the society to improve traffic efficiency instead of spending billions to build new roads for the increasing traffic. Last but not least, service providers must be ready to invest in the infrastructure that can cost a lot of money.

2.2 Terminals Must Fit Services and Supported by Infrastructure

Terminals are required to bring the content of a service to users. The price of terminals as well as the cost for the required infrastructure must be affordable. Car radio is the classical and cheap source for entertainment and information but it does not allow interactive communication. Radio traffic information is not specific for a route and seems not always to be up-to-date. An autonomous navigation system can guide a driver very well how to drive to a certain destination, but it does not know whether the roads are congested.

In Japan a dynamic navigation system called VICS (Vehicle Information Communication System) has been introduced several years ago. To get information about current traffic situation, VICS communicates with the road infrastructure by using a mix of FM broadcast, IR and RF beacons. Such an infrastructure is feasible in Japan but expensive to install in large countries like America or Europe. An IR beacon-based dynamic navigation system has been tested in Germany but it was given up due to the infrastructure cost. On the other hand, there is an infrastructure that exists in nearly all European countries; that is the GSM cellular network. Using this existing infrastructure for telematic services is certainly a sensible approach. In America, starting in year 2001 it is mandatory that a mobile phone that sends an E911 emergency call must be able to provide location information within 125 meters accuracy. This accelerates the integration of GPS and cellular phone and has led to the development of cellular-aided GPS technology. In the following we want to look in more detail at the key components for an intelligent car communication system.

3. Key Components

The basis for an intelligent cellular-based car communication system in Europe is the GSM network and GSM-based services. Other system relevant components are GPS, car multimedia bus, operating system, and hardware platform.

3.1 Cellular Phone Standards and Cellular Phone-Based Data Services

3.1.1 GSM Data Services

Global System for Mobile communication (GSM) is an accepted digital cellular phone standard in Europe and many parts of the world. ERTICO, the European consortium for Intelligent Transport Systems has worked out a strategy for the implementation of GSM-based ITS services [1]. Table 1 shows GSM data services today and in the near future, and how these fit to telematic services that are being introduced by service providers. Today, GSM provides besides data services up to 14.4 kbit/s also Short Message Services. SMS allows the transmission of 160 character strings. Concatenating several SMS can transmit longer messages. For ITS applications SMS is appropriate to transmit traffic information to terminals.

	GSM data services today				Future GSM data services				
Service	voice	data	SMS		USSD	HSCSD	GPRS		
			PTP	CB			PTP	PTM	Anonymous
Emergency Breakdown	+/-	+/-	+	N/A	++	-	++	-	N/A
Traffic Information	-	-	+	++	+	+	+	++	N/A
Floating Car Data	N/A	N/A	+	+	+	-	++	+	++
Route Guidance	-	+	+	-	+	++	++	-	-
Fleet Management	-	+	+	-	+	+	++	++	N/A
Information Services	+	+	+	+	+	++	++	-	N/A
Theft Detection and Recovery	N/A	+/-	+	N/A	+	+/-	++	-	N/A
Remote Diagnostics	N/A	+/-	+	N/A	+	+/-	++	-	N/A

++ very appropriate + appropriate +/- satisfactory - not appropriate

SMS Short Message Services
USSD Unstructured Supplementary Data Services
HSCSD High Speed Circuit Switched Data Services
GPRS General Packet Radio Services
PTP Point to Point
PTM Point to Multipoint

Table 1: GSM data services

In the near future, High Speed Circuit Switched Data Services or General Packet Radio Services allow data rates up to 115 kbit/s. This is nearly twice the speed of ISDN. Downloading a map would not be an issue. Finally, UMTS (Universal Mobile Telecommunication System) the third generation of wireless communication with data rates up to 2 Mbit/s will make possible absolutely new mobile multimedia applications. A good overview about GSM and UMTS as a platform for transport telematic applications is described in [2].

Telematic services or in general wireless communication services require seamless operability between terminals and services. It is crucial that the specific interfaces and functionality are standardized. Currently, there are two proposals for standardization: Global Automotive Telematic Standard (GATS) and Wireless Application Protocol (WAP).

3.1.2 Global Automotive Telematic Standard (GATS)

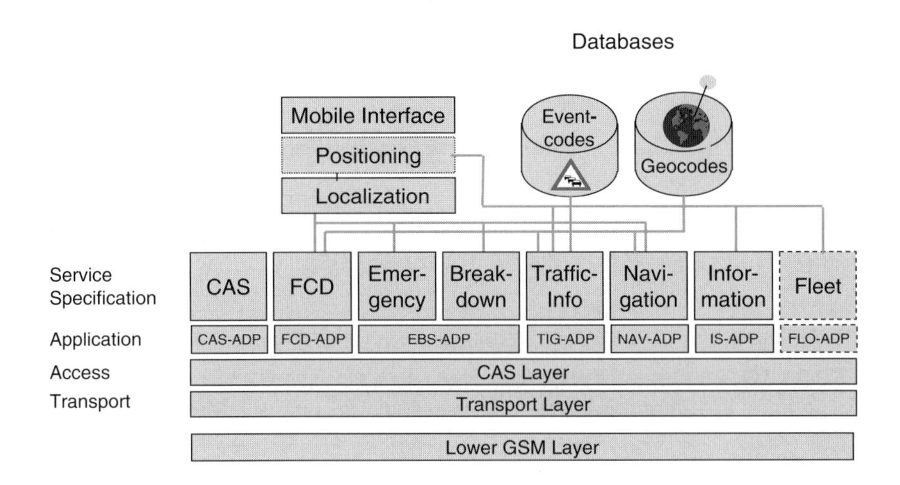

Fig. 3: Global Automotive Telematics Standard

GATS was developed by Mannesmann Autocom and Tegaron, two German telematic service providers. It is being proposed as a European pre-standard (prENV/278) [3] and supported by the GATS Forum.

Fig. 3 shows the telematic services that are covered by GATS. The services supported include Conditional Access and Security (CAS), Floating Car Data (FCD), emergency call, breakdown assistance, traffic information, navigation, general information and fleet management. GATS Floating Car Data is an interesting concept how to use a car that is equipped with a telematic unit as a floating probe for monitoring traffic density. A key cost factor for beacon-based dynamic navigation is the price for installing beacons along the roads. If FCD works well, it can replace a major part of the beacons and would be a very elegant and cost effective solution.

3.1.3 Wireless Application Protocol (WAP)

While GATS is clearly specified as an automotive application, a new Wireless Application Protocol (WAP) has been developed by Ericsson, Nokia, Motorola and Unwired Planet to bring Internet content and advanced services to digital cellular phones and other wireless terminals. Fig. 4 shows the analogy between

Internet and WAP. One of the key requirements for the Wireless Application Environment (WAE) is that it must be optimized for interactive applications that function well on devices with limited capabilities including limited memory, small screen size and restricted input mechanisms. In the meantime all major wireless terminal suppliers have endorsed WAP. WAP is supported by the WAP Forum [4].

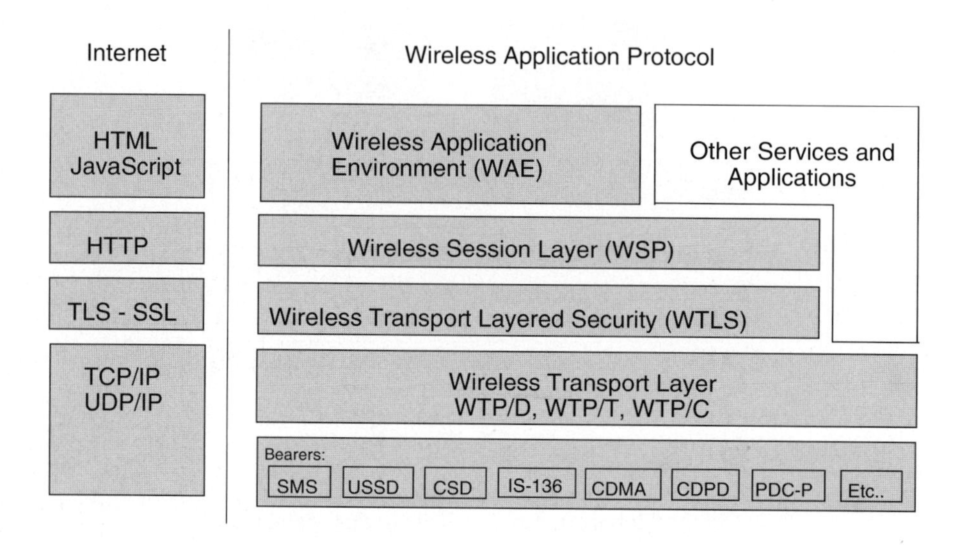

Fig. 4: Wireless Application Protocol and its analogy to Internet

Harmonizing the requirements of GATS and WAP could become a critical point to enable interoperability between various classes of products like telematic unit, navigation system, cellular phone, PDAs, handheld PCs, car computer, etc., and the network.

3.2 GPS Technology

Global Positioning Satellite System (GPS) is widely used in car navigation systems to localize the position of the car. A traditional GPS chip set consists of a RF chip and a digital signal processing chip. To achieve the required precision, a car navigation system needs additional signals from wheel sensors and a gyro. Precise digital maps and sophisticated software are necessary for turn-by-turn route guidance.

An interesting alternative to conventional GPS is a cellular-aided GPS technology developed by Snap Track [5]. When a caller activates Snap Track cellular-aided

GPS technology in the mobile phone, the wireless network sends to the server the approximate location of the mobile, generally the location of the closest cell. The server then informs the mobile which GPS satellites are in its area. The mobile takes a snapshot reading of the GPS signal, calculates its distance from all satellites in view and sends this information back to the server. The server software performs error correction and calculates the caller's latitude, longitude and altitude. Choosing a suitable technology for the application and integrating GPS and cellular phone are possibilities to reduce the cost of the system.

3.3 System Configuration, Car Multimedia Bus

Fig. 5 shows a simplified configuration for a car communication system. At the low-end is the telematic terminal that consists of a GSM phone, a GPS receiver, a gyro and a simple display. There are also products using voice output instead of a display. At the high-end is the navigation system that includes usually a radio with a CD changer, a GPS receiver with digital maps and sophisticated navigation softwares, a board computer, and a high resolution display. A dynamic navigation system needs on top the capability to communicate with the road infrastructure and service providers via a GSM module. A top system can include also modules for TMC, DAB, DVB, PDA, etc. Due to space and cost reasons a common multi-information display must be used for all components. It would be very useful if the system components can communicate with each other via a standard bus.

The problem we have today is that there is no standard for such a car multimedia bus. This could lead to a situation where a dynamic navigation system must include an embedded GSM module although there is already a GSM handset installed in the car. The major players on the market have recognized this problem. At ERTICO, there is a working group to define a Car Multimedia Open Bus Architecture (CMOBA). Apparently, Plastic Optical Fibers (POF) are preferred over copper wires due to lightweight and better EMI.

Among the proposals for a car multimedia bus are D2B, MOST, MML, HiQOS, and Firewire IEEE1394. Besides cost per node, bandwidth is a key requirement for a car multimedia bus. Current implementation of D2B is capable of 5 Mbit/s, MOST 25 Mbit/s, MML and HiQOS 50 Mbit/s, and Firewire from 100 Mbit/s up to 3200 Mbit/s. MOST has been accepted by German car makers. Recently, a new Automotive Multimedia Interface Consortium (AMIC) has been announced during the congress on transportation Convergence 98 in America. AMIC is open only to carmakers. Its members include currently GM, Ford, Mercedes, Chrysler, Renault and Toyota. The goal of AMIC is to create standard hardware and software interfaces for in-vehicle plug-and-play electronics. Finding the right strategy for a car multimedia bus that meets the cost but also allows the bandwidth for future requirements is a real challenge that requires a high degree of consensus.

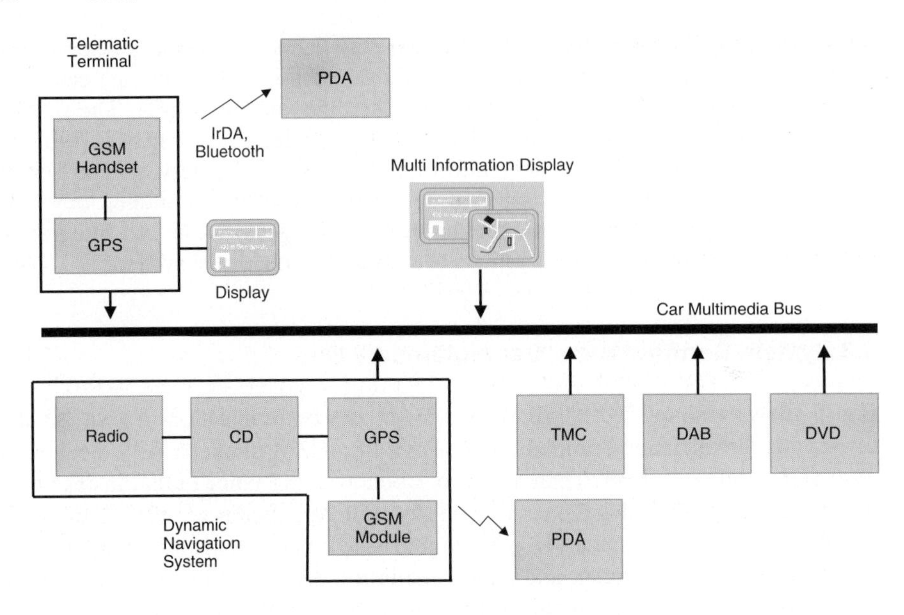

Fig.5: System configuration

Last but not least, wireless connectivity is certainly desirable because it eliminates the cables. Some mobile phones of the latest generation are now equipped with an IrDA interface, a 115 Kbit/s infrared port for connecting the mobile with a notebook or a PDA. Infrared connectivity requires however a line-of-sight. Therefore, it does not always work when the "visual contact" is not good enough. In a similar application, keyless entry has been first implemented with IR technology but now replaced by RF due to better reliability.

Ericsson, IBM, Intel, Nokia and Toshiba announced in May 1998 an initiative for mobile wireless communication called Bluetooth [6]. Bluetooth is a low-power RF technology that operates at 2.4 GHz and allows data rate up to 1 Mbit/s. Bluetooth will enable users to connect their digital cellular phones, notebooks, PDAs, handheld PCs, network access points and other mobile devices via wireless short-range radio links without the line-of-sight restrictions. Users will be able to automatically receive e-mail on their notebook computers via the digital cellular phones in their pockets or synchronize their primary PC with their handheld computer without taking it out of their briefcase. Due to its benefits there is a high chance that Bluetooth will rcplace IrDA.

3.4 Operating System, Human Machine Interface

The success of MSDOS and Windows is a proof that a standard or a de-factor standard for an operating system is a key to develop a mass market for widely accepted softwares. Currently, navigation systems are using real-time operating systems like OS9, QNX or proprietary softwares. On the other hand, Microsoft WinCE and Psion EPOC are popular operating systems for handheld PCs or PDAs. For automotive applications a very compact real-time operating system called OSEK has been proposed as a standard.

Whereas Microsoft WinCE is becoming well accepted as an operating system for handheld PCs that provide a compact version of Microsoft Office applications, Psion EPOC seems to be preferred by terminal suppliers as the operating system for communicators or smart phones. Psion Software, Nokia and Ericsson founded in June 1998 a joint venture called Symbian [7] to support EPOC. On the other hand, Microsoft has proposed the architecture for an Auto-PC. Basically the core of an Auto-PC is a 32-bit microcontroller that can run WinCE as an operating system.

The success of an operating system depends on the fact whether it fulfills the requirements of the applications in which it is being used. In contrast to an office environment, a car communication system has different requirements. These are listed as following:

- Ease of use
- ROM-based, no long boot process
- Low requirements to hardware regarding memory size, screen size and input media
- Real-time, multitasking allowing switch between different applications
- Devices can be switched on and off without loosing data
- 32-bit due to memory addressing space
- Easy to adapt to various devices
- Language independent
- Supported by programmers
- Availability of application softwares

Ease of use and reliability are in fact requirements that should be applicable for all products. In a car environment these features are even more important due to safety reasons. A well designed, ergonomic human machine interface is a must to enable drivers to use complex equipments without distracting them very much from driving. Voice dialing is being introduced as a first application of speech recognition technology in cars. Compared to an office application, speech recognition in cars is by far more difficult due to the noisy environment. Speaker-independent, phoneme-based, noise-robust speech recognition with a vocabulary of several hundred words are currently state-of-the-art. Certainly natural speech understanding would be a desirable solution but there is still a long way to go.

3.5 Hardware Platform

Simply speaking, there are three kinds of microcomputers: microprocessor (MPU), microcontroller (MCU), and digital signal processor (DSP). While MPUs like Intel x86 or Sun Sparc processors are appropriate as a central computing engine for PCs or workstations, MCUs are suitable as embedded controllers for various applications. DSPs, as shown in Fig.4, have on the other hand optimized architecture for extensive high-speed mathematical computation that is required to process signals in daily life. For example, in a mobile phone a DSP must process and compress the data of human voice so that it can be transmitted at a high rate. Besides a fast CPU and dedicated hardware for multiply accumulate, a key feature of a DSP is multiple buses for quick data access.

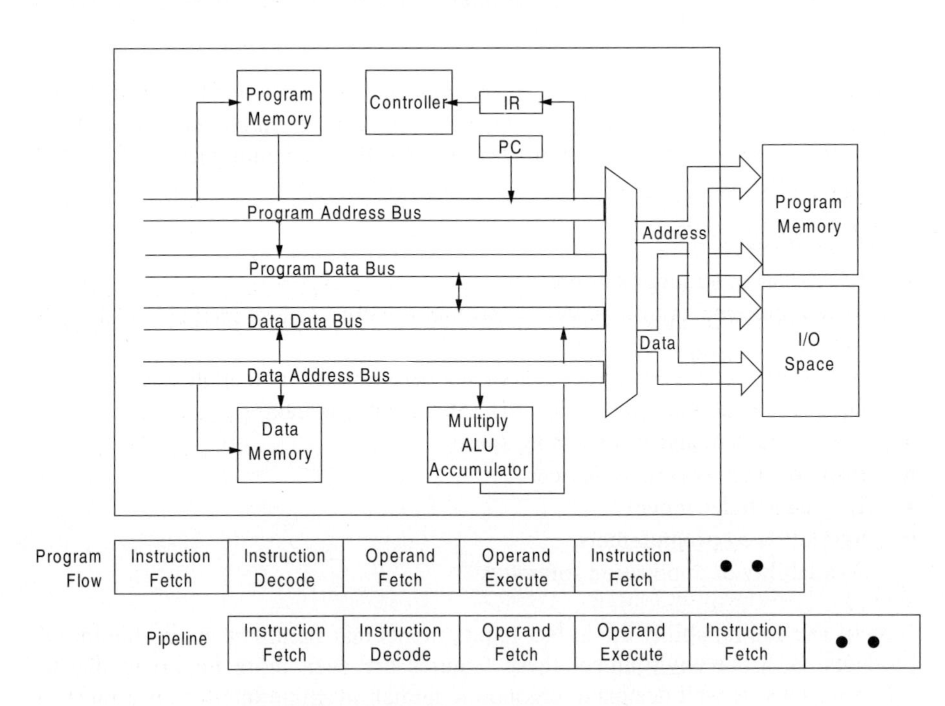

Fig.6: Digital signal processor architecture

Communication and multimedia applications require more and more computing power. New processor architectures using sub-micron processes down to 0.1 micron or less allow billions of instructions per second. Integration of a complete system on one or several chips is a proven strategy to improve cost, performance, and reliability. A system-on-chip requires usually various components: DSP, MCU, RAM, ROM, Flash, RF, and analogue functions. Making such a chip requires besides state-of-the-art processes also sophisticated design tools and technical support. Concurrent hardware software co-design is a desirable methodology to improve design cycle time.

4. Summary

We are seeing a convergence of sophisticated computer and communication technologies in the car. The backbone for an intelligent mobile communication system is the wireless communication platform. In this market, it is not only a question of having the technical capability. In order to succeed a supplier must have the marketing and manufacturing power to produce in high volume to achieve the right cost, performance, and quality. Despite of interest conflicts there is an absolute need for all major players to agree on worldwide standards wherever this is possible to enable a mass-market.

References

[1] ERTICO "GSM based ITS Services - ERTICO strategy for implementation". http://www.ertico.com

[2] Decker P, Salomaki A, Turunnen S: GSM and UMTS as a platform for transport telematics applications. 4th World Congress on Intelligent Transport Systems 1997, Berlin

[3] European standard draft CEN prENV/278/4/3/0015: Road transport and traffic telematics - Traffic and Traveller Information (TTI) - TTI messages via cellular networks

[4] http://www.wapforum.com

[5] http://www.snaptrack.com

[6] http://www.bluetooth.com

[7] http://www.symbian.com

The Application of PC Technology in Automobiles

Jeffrey P. Casazza
Intel Corporation, United States

Synopsis

Computer chips have made their way into virtually every part of the vehicle, from engine control and anti-lock brakes to comfort control and door locks. Today the evolution is continuing with personal computers coming to the automobile. A PC in the car presents exciting new applications for the car occupants and provides manufacturers the benefits of an open, flexible, and cost-effective computing platform.

Intel Corporation has been working with technology leaders in the automotive, computer, consumer electronics and communication industries to develop Intel® Pentium® processor-based computing platforms that offer entertainment, communication, information and navigation in a safe-to-use format. An example design has been developed to aid manufacturers with the challenges of operating a PC in the automobile environment. This example design shows how to create a safe and compact PC appropriate for the car. This paper will present both the concept of integrating the PC in the car and a technical description of an example design.

Keywords: Communication, Entertainment, Information, Navigation, PC-Technology

1. Introduction

A PC in the car provides the driver with many benefits. Navigation programs with real-time traffic information aids the driver in finding the best way to reach a desired destination. No more fumbling for maps or sitting in traffic jams wondering what more efficient route to take. The PC also helps with innocuous functions, like making a phone call and adjusting the radio or temperature controls. Recent negative press coverage criticizing the safety hazards associated with the use of portable phones in car, it has made it clear that the standard desktop PC user interface is not appropriate for the car. A simplified user interface supplemented with voice recognition and speech synthesis allows the driver access to all this information in a hands-free, eyes-free, and safe manner.

Passengers, who don't have the obligation of watching the road, have a plethora of applications available to them. Kids enjoy watching DVD (Digital Versatile Disc) movies, browsing the Internet, playing video games, or finishing their homework. Parents enjoy the children being entertained. When the car is not moving, the driver of the car becomes a passenger and can utilize the connection to the office. An on-line owner's manual helps change the spare tire or find the oil dipstick with instructional videos.

The above applications are not visionary rhetoric. Many of these applications are implemented in cars today. For example, some families already install a TV and VCR into their minivan for holiday drives. Manufactures are producing voice assisted CompactDisc (CD) changers and phones. Navigation units surpassed one million units shipped worldwide. [Williams] Rather than having individual, specialized systems, the future will bring PC's to the car as a multi-purpose system. The PC's open architecture provides numerous benefits. Some navigation systems offer voice- controlled mobile phones, Japanese systems offer the adding of games, and karaoke. Several use the UI (user-interface) to control heat/cooling functions. As applications become more complicated, the investment required in a proprietary system becomes significant. Development resources can be used to bring value to a proven hardware and software platform rather than developing a specialized system. There is also the advantage of using the enormous base of existing building blocks. Standardization will allow the use of existing software, both applications and other core components like drivers and APIs.

Applying PC technology to the automobile presents several design challenges. Ease of use and safety must be designed in from the start. The architecture must provide standardization to keep the cost low and provide flexibility to allow product differentiation.

This paper discusses how a PC can be adapted for use in automobiles, the architecture of the system and how it interacts with the subsystems, key design challenges, and an example development system.

2. Definition of subsystems

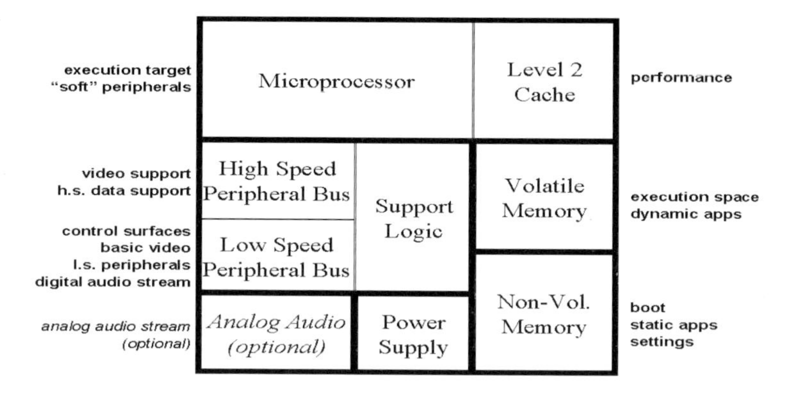

Fig. 1: Block Diagram of CarPC Computing Core

All CarPC features are supported through a general-purpose computing core. Features are added to the system through the attachment of individual subsystems. For example, control of the climate and other comfort controls is supported by including car-bus interface hardware with the obligatory support software. The modular nature of the system allows flexibility in designing feature packages.

2.1 General purpose core

The general-purpose personal computing core is the main subsystem and the flexible "brains" of the CarPC. It contains the microprocessor that executes the system software. Since the microprocessor is based on Intel architecture a wide variety of development tools and off-the-shelf software building blocks are available to the developer. Also included in the core is a level-2 cache, which improves the overall performance. The cache is implemented as SRAM, and will vary in size depending on the required performance of the final product. Approximately a 20% increase in performance is realized by using cache versus a system without cache. The array size will vary from 0-512KB.

The volatile memory, an 8-64MB DRAM array, is relatively low-cost and high-speed providing the necessary memory space for executing software. Software added in the field can be stored to the volatile area, providing support for user upgrades.

A 4-32MB flash array is used for non-volatile memory storage. This space is used for storing the operating system and built-in applications. It is from this space that the system "boots." The software is stored in a compressed format and is decompressed and shadowed to DRAM for execution. Theoretically, the software could execute directly from the non-volatile memory. However, flash access times are slower than DRAM. The extra wait states cause a high-speed microprocessor to stall, so that its performance would be equivalent to a much slower microprocessor. The flash array also provides a non-volatile area for storing system and personal settings.

The power supply provides the necessary power conditioning and regulation support for the system electronics. The support logic provides the glue for connecting the different components. It also contains the peripheral buses for connecting the additional subsystems. A low-speed path is provided to allow the interface to the control surfaces, that is, the buttons, knobs, and other mechanisms for the user to interact with the system.

The bus also supports low-end display functions such as a simple Liquid Crystal Display (LCD). The digital audio stream is supported using the low-speed bus for the gateway to and from the PC. Finally, any additional peripherals that require low-bandwidth connections attach to the bus. Examples include joysticks for games, car-bus gateways, telephony, and GPS. The bus could be implemented through one of several protocols, such as USB, RS-232, ITS-DB, or some other <15Mbps bus.

A high-speed bus is provided to connect data intensive peripherals, such as high-resolution monitors and CD/DVD drives. This bus could be a local bus like PCI or EIDE for simple in-the-box connections or longer distance buses, such as, IEEE-1394, MOST*, D2B*, or other fiber-based protocols.

For supporting the older analog audio model some additional circuitry (e.g. CoDec) is required. The audio acceleration functions are provided through software running on the microprocessor.

2.2 Audio

Audio is integral to any car multimedia system. Today there are two basic ways to implement PC audio: an audio accelerator model and newer digital audio stream paradigm.

The PC audio accelerator model has a CoDec that has analog inputs and outputs. This connects to the accelerator component via an AC97 link. The device resides on a PC bus such as ISA or PCI. Commands are sent to the accelerator, which performs any required mixing and signal processing. Moving forward, future PC's will have the signal processing, mixing, and other audio accelerator functions performed in software running on the host processor. This reduces the audio subsystem hardware to just an inexpensive CoDec.

USB also supports audio. A digital audio stream can be transferred over USB using its isochronous mode. Since the audio is digital, it allows for control and enhancement to be performed by host software. It also reduces the wiring requirements by using a serial bus structure instead of individual analog wires. Designs appear to be trending to a digital solution in the future. One benefit is that a USB audio device can not only transfer the audio data, but command data as well, allowing the central computing electronics to take control of the audio device. This would be useful in a scaleable radio scenario. A radio with a simple display and a USB port could be installed in a car stand-alone. If the user wants to add functionality, a dealer-installed or after-market CarPC product could be added. The CarPC can control the radio functions and use its display to realize new functions, such a navigation or telephony.
USB is not the only bus to support digital audio. 1394, MOST*, D2B*, and other can be used to implement audio streams. All of these buses have hardware implementations available for the PC, typically via the PCI bus.

2.3 User Interface

The user interface of the CarPC is the area of most interest, as it is the most visible part of the entire system. The basis of consumer acceptance will be based more on the user interface than any other features, except possibly the cost. To implement an effective user interface, a display and an assortment of controls is presented to the user.

The display is an LCD with a typical resolution similar to basic computers, 640x480. The size will be decided by the usage and cost model, but will average 5-6 inches for most systems. A graphics controller with an integrated frame buffer drives the display. To reduce cost a single integrated solution will be used. Standard LCD or VGA outputs have a limited range, so to remotely locate the display from the main electronics some technique must be employed to condition the signals for transport (e.g. PanelLink*).

Control of the system uses various buttons and knobs. These surfaces provide the user with simple tactile controls. Using a USB microcontroller allows the information from the buttons and knobs to be easily converted to a standard PC interface. In addition to the tactile interface, a speech recognition engine running on the host processor provides a supplemental method for controlling operations. Audio feedback and text-to-speech technology provides the user with a menu of options and acknowledgement of the command.

As the user interface is so important to the system style, its implementation allows for OEM's to provide differentiation in their products.

2.4 Car comfort control

Having a screen available for a user interface and a general-purpose computer for processing, new car comfort control implementations are possible or at least centralizing of existing functions. By providing a gateway to non-mission critical car electronics, the CarPC displays car information, such as odometer, engine temperature, and diagnostic information. It also provides the interface for controlling car comfort controls like the temperature control system, seat position, and radio. The graphical user interface supplemented with speech recognition allows for easy-to-use implementations. The non-volatile storage provides a means of saving custom configurations, for example the seat and radio settings for different drivers of the same car.

To implement these features, the computing core needs a connection mechanism to the car bus. Since car-buses typically have low data rates the most obvious portal to the PC is USB (RS-232 is only recommended for interim solutions as it will be removed from the standard PC platform in the future). A USB microcontroller can be connected to a car bus controller to create a car bus gateway. This gateway provides a means of the car electronics and the PC to interact.

2.5 Telephony

People like to use their telephones in the car. Connecting them to a CarPC system can make them easier and safer to use. The user interface of the computer with its large buttons and speech recognition software can improve the car phone experience. Messaging and fax services may also be provided. To offer all of this functionality the CarPC and telephone must be connected. Although the phone electronics could be physically embedded into the system electronics, cost and user-performance considerations favor a docking solution. The recommended connection mechanism is through a serial port such as USB.

A docked USB phone would allow the CarPC to control the phone dialing functions leveraging the CarPC's user interface. Voice data can be passed between the incoming call to the computer in digital form and on to the car's audio system. Likewise, the microphone for the CarPC can be used for the phones voice pick-up. Because the information is digital, data calls can be supported with no modem. This reduces the cost of the system. Additionally, USB provides plug-and-play support allowing the computer system to detect when a phone is present.

For legacy support of older analog phones, a soft modem can be implemented on software on the host processor. A CoDec is required.

2.6 Navigation

Accurate car navigation systems utilize two techniques for determining the vehicle position: global positioning system (GPS) and dead reckoning. The dead reckoning technique uses direction and velocity to calculate the position from a known point. It complements the GPS when buildings or tree cover interfere with its normal operation. It also improves the overall accuracy of the system which would be otherwise limited to the GPS's 100 meter accuracy due to a government introduced error called selective availability (SA). As the technology matures, the positional electronics' cost and complexity is reduced. GPS chipsets allow a GPS receiver to be implemented for about $30. With an optional gyro and A/D converter, dead reckoning can be implemented without the cost of an additional microcontroller. [Turley] Most receivers interface with the main computer via a serial link, such as RS-232. Future GPS sub-system may transfer some of the general purpose processing to the host processor to further reduce the cost of the system. As Universal Serial Bus devices become available, they will supplant their older RS-232 counterparts.

Besides the positional equipment, the rest of the navigation processing such as the map matching, route calculation, and user interface are accomplished using the main general purpose computing core. Input follows the same paradigm as other car functions, with the graphic display and surface controls, supplemented with voice commands, providing the user interface. A CarPC can be modified, for a small incremental cost, to provide state-of-the-art navigation functionality.

2.7 Entertainment

If the system is to support DVD movies, the compute-intensive MPEGII decode must be supported. Initial systems may use dedicated hardware for the MPEGII decode, however, in the near future, 300+ MIP's processors will be readily available and cost effective. Such processors can decode the movies without the additional cost of dedicated hardware.

For systems that require sophisticated game support, the graphics controller should include 3-D acceleration. This will allow the gaming effects to meet the high expectations of today's computer gamers.

3. Design challenges

Designing automotive electronics is more challenging than basic consumer electronics devices. The car environment has similar cost constraints of a successful consumer device, but with added problems such as harsh operating environment, small form-factor, and occupant safety considerations. Any applications of PC technology into the car must address these issues before successful deployment.

3.1 Automotive environment

PC components are solid state and can be made available to extended temperature and qualification levels with cooperation of vendors. Microprocessors, chipsets, and non-volatile memory are available in extended temperature. Other support components such as DRAM, clock, and power supply may not be readily available in extended temperature. Cooperation between vendors and OEMs are required to ensure the proper components are available. The weak links to surviving the harsh environments of the automobile are not the solid state components, but rather the display and rotating media.

The most common display proposed for automotive applications is the liquid crystal displays (LCD's). Unfortunately, LCD's do not have a wide enough operating temperature range for the car. Early systems work around this limitation by providing a controlled environment for display, usually in the form of a heating apparatus to warm the display in cold temperatures. This is an effective, albeit costly solution. A new technology called field emission display (FED) promises a wider temperature range with the additional benefits of brightness approaching conventional cathode ray tube (CRT) displays. Several manufacturers are currently starting production and some are specifically targeting in car applications. [Siano]

The other weak link in the system is rotating media, CD's, DVD's, and Hard Disk Drives (HDD's) all have difficulty surviving in an automotive environment. Automotive CD players are the most mature and have been successfully deployed in millions of autos. Their susceptibility to shock and vibration has been overcome by providing damping mechanisms and data buffers. These buffers "read ahead" in the audio stream. In case of an error, the information is read from the buffer, while the mechanism realigns. Unfortunately, this solution doesn't work well with non-streaming data, like PC storage data. Instead, the drive reports an error and the data read must be retried. The effect is a performance penalty, but operation in a bumpy environment.

DVD's are very similar to CD's and promise to support the same operating environment as the technology matures. Some manufacturers are discussing selling DVD drives to car applications.

Hard disks also present design challenges in a car environment. Currently, there are at least two navigation systems in production vehicles that use HDD as mass storage device. [Siano] Surprisingly, the shock and the vibration are not the main limitation. Sophisticated mechanical systems and error detection mechanisms allow HDD's to operating in relatively high-g environments. The primary drawback is the drive's narrow temperature range and susceptibility to moisture condensation and freezing on the media. Besides controlling the temperature to the drive, like early LCD application, another solution is to use a solid state based system. Operating systems and applications suites with a small memory footprint allow system implementations of 8-16MB of flash; a reasonable cost target for

CarPC systems. Supplemented with the large data capacity CD or DVD drive such systems rival the functionality of HDD-based systems.

3.2 Safety

One of the initial concerns to those first exposed to the CarPC concept is safety. Recently, the use of cellular phones has come under fire as a safety hazard. A number of news articles have been published and government regulatory agencies are starting to investigate the issue. [Walters] Newcomers to the CarPC concept view the operation of such a device as a sizable distraction. However, one of the key design goals of the system must be safety. Proper design of the user interface and control of the material presented to the driver are important criteria for purposing a non-obtrusive system.

The user interface is the most important aspect to a safe design. The controls must be easy for the driver to operate. Any buttons, knobs, or other control surfaces, must be easily accessed by the driver and cause no distraction to normal operating duties. In addition to the control of the system, the presentation of information in a visual form must be well engineered. A simple, easy to comprehend visual design must include large text and easy-to-read information. Screen commands should be intuitive and menus should not be more than two levels. Human factor studies on navigation systems provide a foundation of good CarPC design. [Noy] [Fukuda]

Speech technology to supplement the mechanical and visual interface can further reduce the hazard. Voice recognition software is becoming accurate and inexpensive. In particular, several companies are demonstrating command recognition software that can recognize with high accuracy a limited vocabulary, independent of the speaker. Coupled with speech synthesis to provide audible feedback and menu options, most PC-based functions that the driver needs to perform while driving can be performed eyes and hands-free. This design approaches a first level goal of being as safe to use as talking with a passenger.

A final implementation technique to create a safe design, is the control of information to the driver. A mechanism is provided to determine if the car is in motion. The level of detail and type of information presented changes depending on whether the vehicle is in motion. For example, a movie should not be viewable by the driver while they are operating the car, but may be acceptable if the car is stopped and/or parked. Likewise, a detailed map may be removed from the screen and replaced with at-a-glance directional icons and indicators once the car becomes in motion.

3.3 Cost

No discussion of a deployable commercial automotive system would be complete without a discussion of cost. For mass-market acceptance, the price to feature relationship must be appropriate. Building around a general-purpose PC core allows cost-effective implementations and low-cost building blocks. Borrowing technology trends from the value priced Basic PC market segment; integration will provide the key to the low-cost solution.

One design philosophy is to use specialized, dedicated hardware to implement each of the functions. This approach has the unfortunate effect of duplicating components. For example, a simple navigation system may be two CPU subsystems in the enclosure: one for user interface control and the other for GPS calculations. Each of these subsystems has a microprocessor, RAM, and ROM. In addition, application specific hardware reduces the flexibility of the system. To add functionality to the system, for applications like games, a third subsystem needs to be added. Using a general purpose PC paradigm, functionality can be added by including more software applications and increasing the performance of the CPU and the size of the memory array. This paradigm allows for the development of a family of system services different with price point and feature sets.

To achieve a cost-effective solution, a general-purpose computer core should be used. The system will be flexible to service a variety of feature sets. Using widely available and mature commercial hardware and software building blocks provides a low-cost basis for fielding CarPC systems. Off-the-shelf building blocks also reduce the investment required to develop custom applications and hardware designs. Differentiation is provided by tailoring the specific features targeted and through developing a customized user interface including the mechanical styling and graphical presentation theme.

3.4 Packaging

The automotive environment poses challenges to the packaging of the system components. The form-factor of the enclosure must be appropriate for the application. One target is for an in-dash unit that replaces the audio system and comfort controls. Another method is to partition the system, mounting the system electronics remotely (e.g. trunk/boot) and installing only the user interface components in the dashboard. Regardless of the installation methodology, the overall size of the unit should be minimized. The form-factor is not a challenge. There are small form-factor computers readily available on the market that implement and entire PC, including the LCD, in less than 53in3.[Toshiba]

A greater design challenge is the development of an appropriate thermal solution. A typical notebook PC dissipates 15 to 20 Watts using only passive thermal techniques. Between 2 to 8 Watts of this budget are allocated to the microprocessor. [Intel] (It's worth noting that this thermal range provides for a 100-300 MIP processor, far exceeding the computational power in today's navigation systems)

The operating environment of a notebook computer, however, is in a narrower ambient temperature range. Components themselves are qualified to an extended temperature range to improve the thermal range. In addition to more rugged components, an active thermal solution may be required to achieve the desired automotive temperature range. Driven in part by the notebook computer industry, new high-reliability fans are available. These bearing-less fans have a mean time between failures of 100,000 hours. Most designs will target a passive thermal solution. In this case, the enclosure must be designed to remove the total system heat. A low thermal resistance heat path must be provided to move the heat to the removal mechanism (e.g. enclosure heat sink). Examples of extremely high heat removal systems are the enclosure of high-power amplifiers. Although the CarPC solution will not need to be quite as elaborate, consideration must be given to ensure proper thermal design.

3.5 Appliance/Radio-like operation

A final design consideration is that the device must operate like an appliance such as a television or a car radio. Both turn on very quickly and are simple to use. They are robust, in the sense that no system crashes occur. To accomplish these design criteria new software technology must be applied.

Initiatives like Instant-On (OnNow) provide suspend and resume functionality that does away with the historic and notoriously slow boot time of a personal computer. The user-model displayed by the generation of hand-held PC companions have applications running from flash memory and DRAM. The power button merely suspends operation while the battery supplies power to refresh the contents of memory. In a true reset power on condition, both of these techniques demonstrate a slightly long response time.

4. Development platform example

A development platform has been designed to aid manufacturers in applying PC technology to the automobile. The platform is meant to be three systems in one, serving as a reference design, development system, and a demonstration vehicle. Given the three conflicting design goals, the system is purposely not optimized for any one goal. A large feature set is supported on the development system to provide developers flexibility.

Fig. 2: CarPC Development System

Development System Features:
- Single screen UI with button and knob controls

- Remotely locatable display
- Software shell utilizing voice recognition
- Ability to run off-the-shelf PC applications
- Support several operating systems including Windows* 95 and Windows CE
- Internal space reserved for hardware and peripheral development
- Form factor approaching product sizes
- Nominal automotive operating specifications
- Incorporates GPS functionality
- Provides four RS-232 and five USB serial port
- Infra-red receiver ready
- Support both HDD based and Flash based implementations
- CDROM audio/data drive
- Support standard car audio devices

The system is implemented with an Intel Pentium Processor with MMX technology running at 266MHz. The system can be adapted to support a Pentium II Processor. The mobile 430TX chipset is used to provide advance power management functionality and thermal management capability. To support the requirement of PCI hardware development areas, the main electronics are divided across two double-side printed circuit boards.

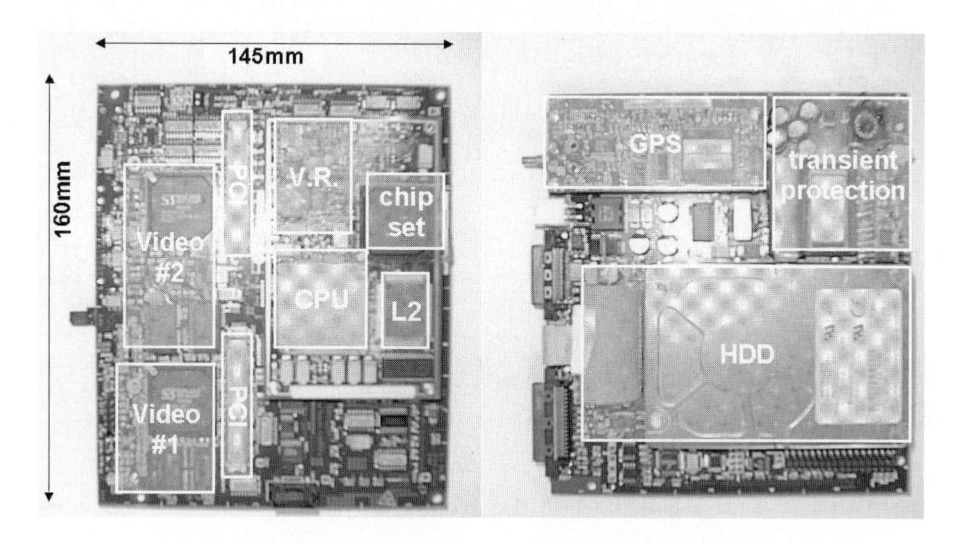

Fig. 3: Two PCBs comprise the CarPC Electronics.
(Some functions on the other side of the boards are not visible)

4.1 Development System

A number of features are provided to support both hardware and software development. Four RS-232 ports and five USB connectors are provided to allow external peripherals to be connected. Existing RS-232 equipment, such as GPS units, can be connected to the system using these ports. Internal connectors have also been provided to allow the development of PCI- or USB-based functions. Using these areas, additional functionality such as CAN support or MPEG2 decoding can be added to the system without increasing the form factor. Similarly, the system is a "software target" for evaluating car applications. The display enclosure provides a software developer with a series of buttons and knobs to develop and target more auto-friendly software. Because development usually occurs indoors, the system can be set up on a laboratory bench. Power can be supplied through an adapted PC ATX power supply. Additional switches have been added to simulate in car events.

4.2 Demonstration Vehicle

Unlike its desktop brethren, the CarPC development kit was designed to be small enough to be mounted in the vehicle. The separate user interface enclosure contains the screen, knobs, buttons, and USB connectors required to support an in-car man-machine interface. The user interface software allows operation from either the control surfaces or using speech recognition. No keyboard or mouse is required for the driver-centric applications. The core electronic enclosures are DIN-size to allow mounting in the space constrained car platform. The power supply supports connection to the 12-Volt car electrical system. Transient suppression is provided to prevent damage from the noisy car power distribution system. An external event line is provided to implement a system wake up from a car device, such as the ignition/accessory switch. The standard RCA jacks allow the system to be connected to standard car amplifiers and radios.

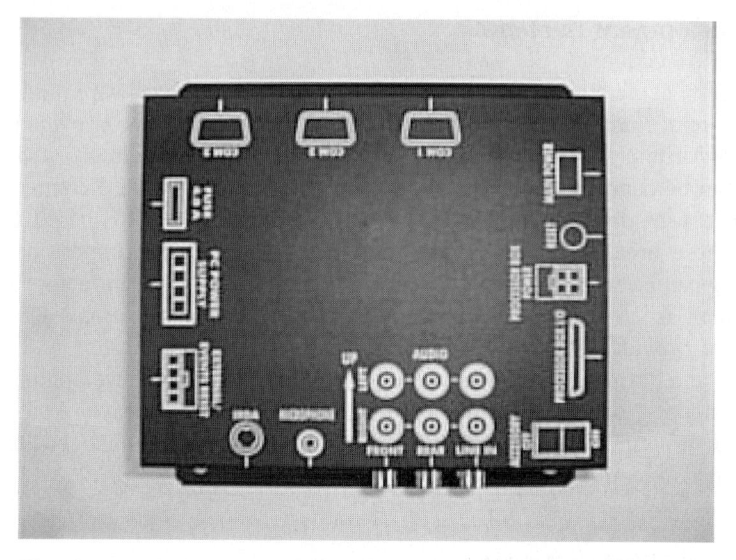

Fig. 4: Dongle box provides lab bench development connectivity

4.3 Reference Design

Both hardware and software example implementations are provided to assist manufacturers. The core PC architecture is an example that can be referenced for similar designs. The software provides an example of how different the user interface is in a car versus on a desktop. The software controls are implemented using ActiveX* to demonstrate portability. Since the platform is not targeted as a specific product, it is not optimized for production. Known limitations include a less than ideal thermal example and the utilization of an EIDE CDROM drive. Actual systems would require a different bus, such as 1394, in order to support a remotely located CD-ROM. Production systems would need a more robust thermal solution to ensure components maintain specified operating temperatures.

5. Conclusion

The PC is the ubiquitous computing platform for millions of consumers. The industry momentum behind PC technology ensures a constant stream of cost-effective technology developments. In addition, it provides developers with a familiar development platform. Using PC technology in a car allows the consolidation of numerous communications, information, and entertainment functions into a general-purpose computing core. Safe operation of the computer is enabled by the intelligent design of the user interface and assisted with voice recognition technology. With these advantages, it will be no surprise to regularly see CarPC's in automobiles in the not-to-distant future.

6. Acknowledgements

Special thanks are due to Intel's Connected Car PC design team including: Surendra Burman, Dan Carey, Mark Eidson, Rick Feightner, Bruce Fleming, Bill Harris, Kevin Jayne, Greg Scott, John Merrill, Lee Noehring, My-Hahn Nguyen, Mark Sullivan, and Paul Zyskowski. Their dedication and innovation created a truly unique PC development platform. And for his leadership and vision of bringing PC's to car, thanks to Patrick Johnson.

7. References

Avalos, Jose: 'Intel Connected Car PC Platform Development Implementation Case Study.' Proceedings of 31st International Symposium on Automotive Technology and Automation, June 1998.

Casazza, Jeffrey P. and Kumakura, Tsuyoshi: 'Designing PC's for Use In Automobiles.' Proceedings of the High-Performance System Design Conference, January 1998.

Intel Corporation: 'Mobile Pentium Processor with MMX Technology on 0.25 Micron.'.

MacLellan, Andrew; "FED tech ready for market move" Electronic News May 19, 1997 v43 no2168

Schlott, Stefan: Vehicle Navigation: Route planning, positioning and route guidance. BMW/Philips, 1997.

Siano, Joseph: 'Exploring the Maze of New-Car Navigation Systems.' The New York Times, April 26, 1998.

Toshiba Computer Systems Division: 'Libretto 70CT Detailed Specificaitons.'

Turley, Jim: 'SGS-Thompson Crams GPS Into Two Chips.' Microprocessor Report, Microdesign Resources March 30, 1998, p.10-11.

Walter, Barbara: 'Hidden Risks of Using Car Phones.' ABC 20/20 Transcripts, August 22, 1997.

Williams, Michael: 'Driver Information, Communication, and Entertainment.' Dataquest, June 1997.

---

* Third party brands and names are the property of their respective owners.

** USB and ITS-DB are abbreviations for Universal Serial Bus and Intelligent Transportation Systems Data Bus, respectively. MOST stands for Multimedia and Control Networking Technology.

* Third party brands and names are the property of their respective owners.

* Third party brands and names are the property of their respective owners.

* Third party brands and names are the property of their respective owners.

Advanced microphones for Handsfree Communication and Voice Recognition

Robert Frodl
Ruf Electronics GmbH
Bahnhofstr. 26-28
D-85635 Höhenkirchen

Abstract: The booming cellular phone industry is the pacemaker for a new statement: Voice processing technologies are key elements in modern automotive communication systems. Cellular phones, voice links to car operating systems, and in-car computing networks will require advanced microphones to achieve the desired performance levels. These active microphones allow the driver to communicate in a handsfree mode with his communication partner, regardless whether human or machine, adding safety and comfort to driving a car.

Directivity, unsensitivity to wind, frequency response, vibration isolation and other parameters need to be optimized for the human voice within the noise and sound floor levels of today´s driving environment. Both theoretical calculations and on-vehicle testing are necessary to position the microphone perfectly to deal with the complex acoustic patterns found inside a car. New approaches for noise cancelling and improvement of intelligibility are discussed in a new design.

Keywords: Handsfree Systems, Microphones, Voice Links

1. Introduction to Automotive Microphones

Automotive microphones have to pick up sound in an automotive environment. Sound waves are mainly described by their frequency or frequency bandwith and their sound level or sound level range.

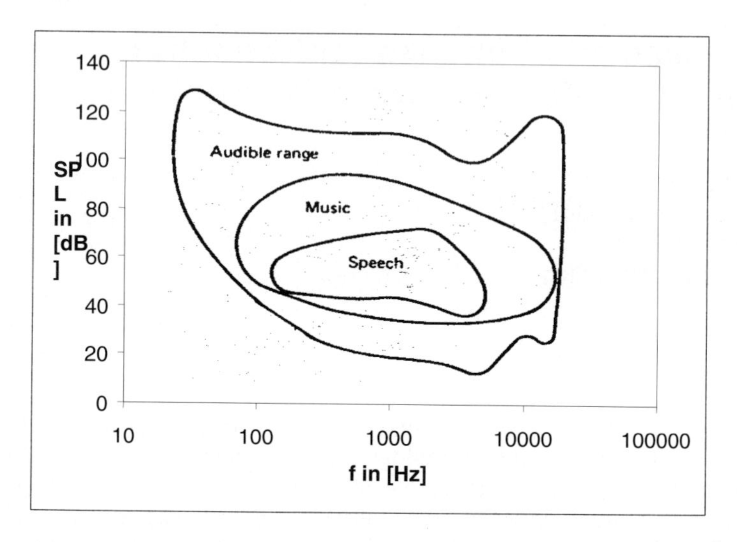

Fig. 1.1: Operating conditions for automotive microphones regarding frequency range and sound pressure level (SPL) range[1]

Practical applications will be in an area from audible signals beginning at 10Hz up to 100kHz which is known as ultrasonic. The according sound pressure level (SPL in dB) reaches from 0 to 130 dB SPL. Microphones for cellular phone purposes have to work in a limited bandwith within the speech area (Figure 1.1). Applications which utilize ultrasonic technology have to make sure that they do not disturb animals like dogs and cats because they may be inside or beside a car. This is one of the reasons for the lower frequency limit of about 40kHz for ultrasonic systems.

Out of the numerous applications (Figure 1.2) this paper will concentrate on inside car voice microphones.

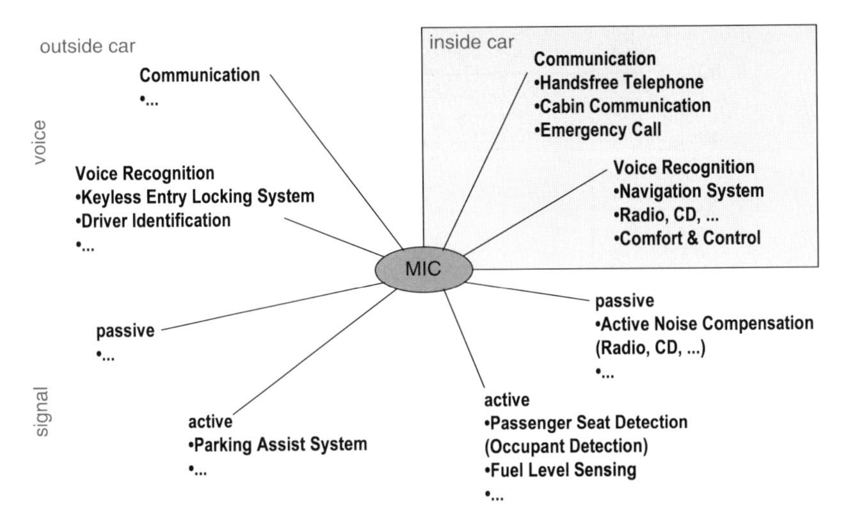

Fig. 1.2[2] **:** Automotive microphone applications sorted by microphone location and type of signal

2. State of the Art Microphones for Handsfree Communication

2.1 New Systems need New Microphones

Driving force for the growing interest in active microphones for handsfree communication purposes is the booming cellular telephone industry[3] (Figure 2.1, 2.2) as well as recent achievements in voice recognition algorithms. A similar trend can be watched in the take rate of GPS based navigation computers. Recent achievements in voice recognition technology allow to command navigation systems as well as radio, CD and comfort controls.

Several investigations show that higher crash risk occurs when a driver communicates with a handheld cellular phone while driving a car. Several administrations worldwide prohibit the usage of such a handheld cellular phones during driving or recommend at least a handsfree system[3]. A handsfree system allows for speaking while keeping the hands on the steering wheel. This improves safety as well as comfort.

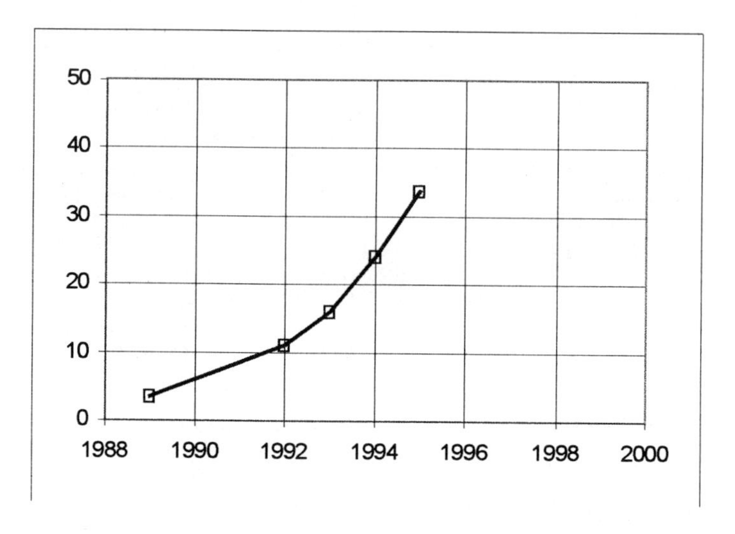

Fig. 2.1: Cellular phones in the USA (in million units)[3]

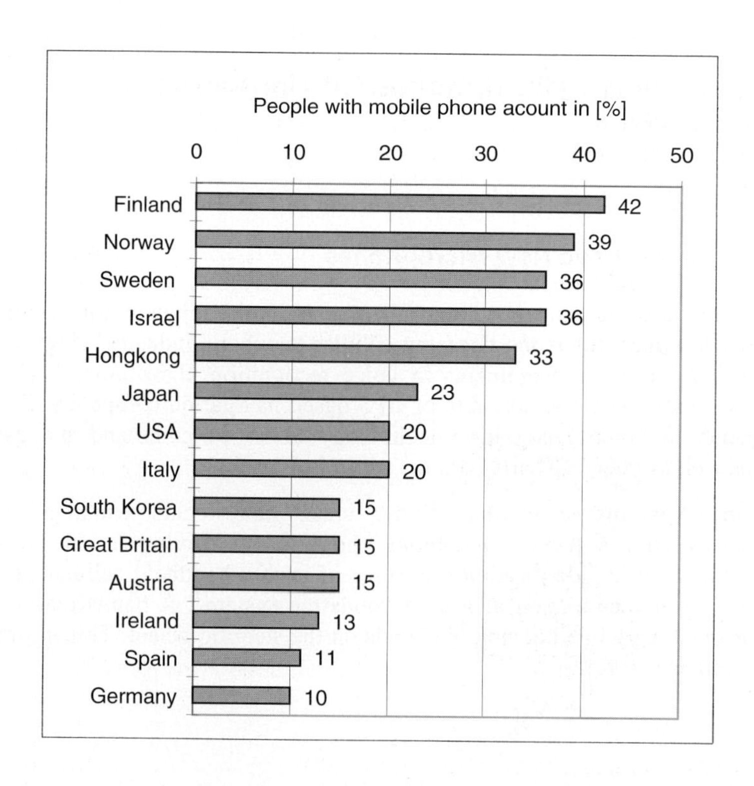

Fig. 2.2: Mobile phone accounts in 1997[4]

The first generation of voice recognizers needed exact pronounced single commands. The new generation of algorithms, which are already under development, will allow the interpretation of fluently spoken text. The acoustic performance of microphones has to keep pace with this development because sophisticated speech recognition functions need a clear signal of the spoken words while the signal to noise ratio of a speaker's voice in a driving car is limited.

Scenario of fully equipped vehicles

Fully equipped vehicles do have up to 5 different microphones to support all systems[5]:

1. Mobile phones
2. Voice recognition for command and control purposes like GPS
3. Emergency systems via satellite or radio communication
4. Sound system microphone for measuring ambient noise for noise leveling (for example Dynamic Range Optimizer DRO[6])
5. Radio communication for taxi drivers and government services like police

It is evident that one active single microphone with an appropriate standardized interface to all systems would be of valuable benefit. At least for mobile phones there are standardization groups in Europe[7,8] as well as in the United States. For full systems different approaches are discussed by some OEM´s (for example with optical bus systems[5]) and system suppliers[9].

State of the art microphones

State of the art microphones are costumized to every single application for every single car model to achieve the desired performance for the system. The process of costumization has to take into account a list of parameters beginning with the right location in the car, ending with influences of the interior design.

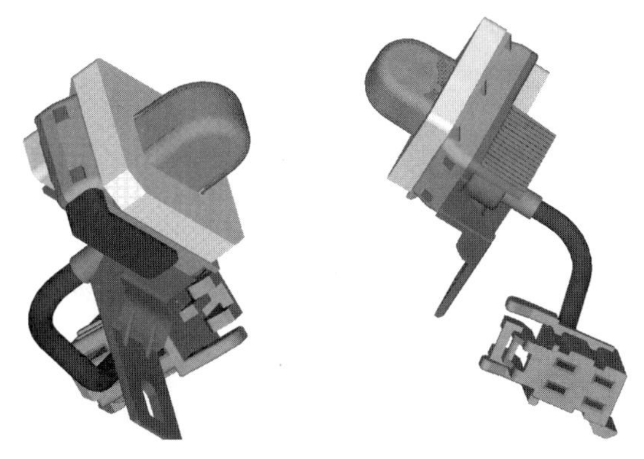

Fig. 2.3[2]: Ruf Electronics state of the art microphone for an american luxury car, in series production since June 98 (mounted on steering column)

Fig. 2.4: Ruf Electronics state of the art microphone with push button for emergency systems for clip on mounting (mounted on A-pillar)

Figures 2.3, 2.4 demonstrate state of the art active microphones in series production.

In the following paragraphs we will discuss the most important parameters which have to be considered in order to achieve sufficient solutions.

2.2 System approach for active microphones

In a handsfree cellular operating system there are different acoustic pathes which interfere with each other. The microphone design has to find solutions to enhance the path speaker to microphone (Figure 2.5) and to supress disturbing pathes from noise sources to microphone (Figure 2.6, 2.7, 2.8).

Especially the interference from loudspeaker to microphone is an important issue. As long as the handsfree system operates in a semiduplex mode the loudspeaker may only be a noise source, i.e. if the radio is on. At the moment when the handsfree unit has to operate in duplex mode, to which customers are used to by standard phones, things become difficult.

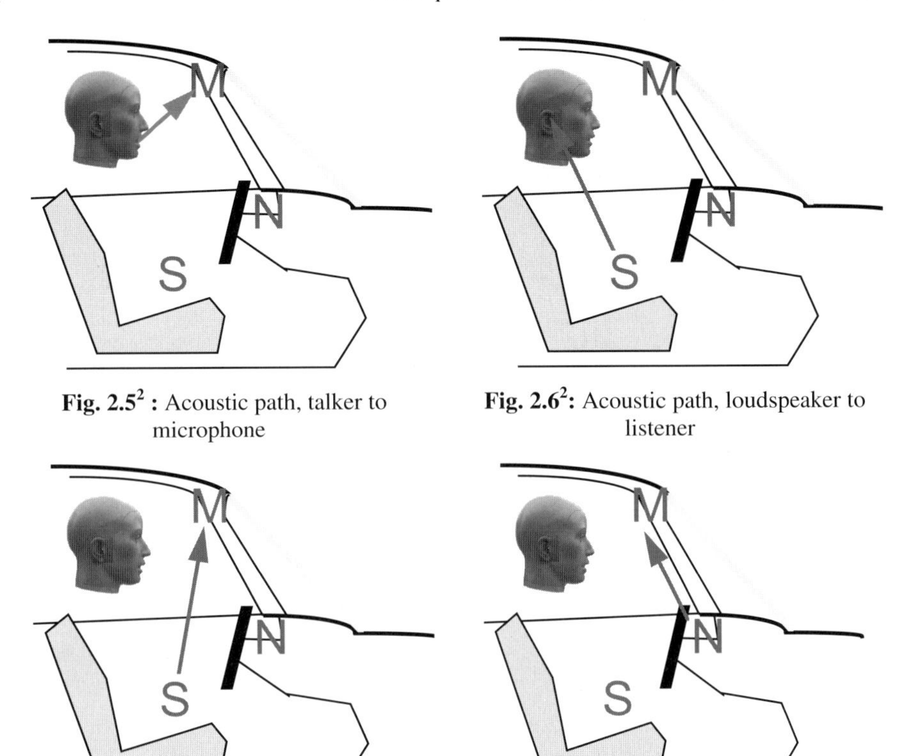

Fig. 2.5[2] **:** Acoustic path, talker to microphone

Fig. 2.6[2]**:** Acoustic path, loudspeaker to listener

Fig. 2.7[2]**:** Acoustic path, loudspeaker to microphone

Fig. 2.8[2]**:** Acoustic path, noise sources to microphone

If both parties communicate in handsfree mode, the signal being picked up by the microphone of party 1 will be transmitted to the loudspeaker of party 2. The amplified signal will be picked up by the microphone of party 2 and transmitted back to loudspeaker of party 1. Party 1 would hear its own echo and additional feedback signals throughout the conversation.

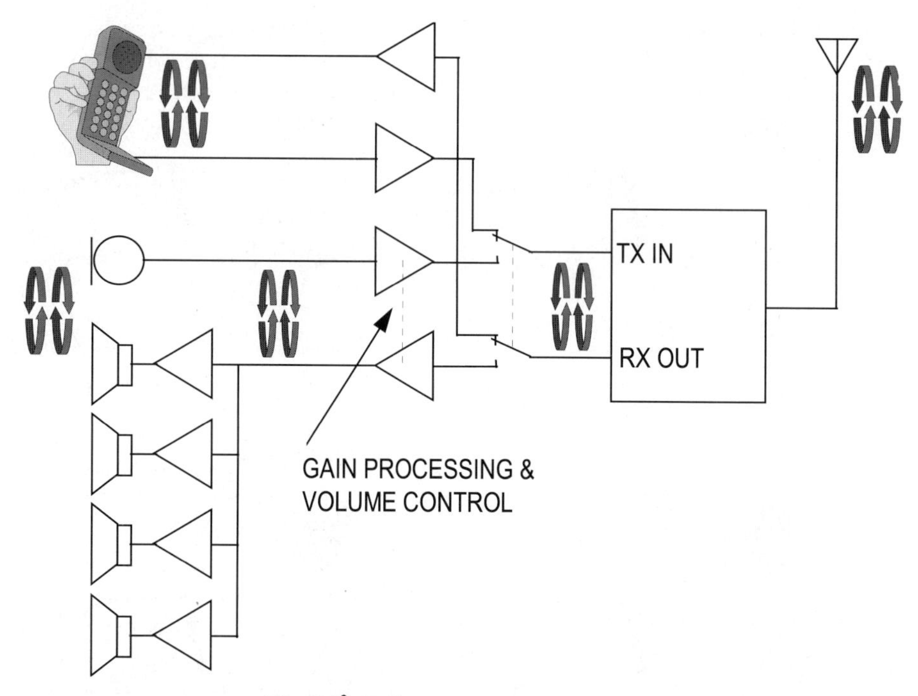

Fig. 2.9[2]: Talker Listener interference

Current systems use electronic gain and volume controls to supress acoustic echo. These controls realize a semiduplex mode by switching the participant on or off in dependency whether they speak or not. Or in other words: who speaks loudest will be heard.

For full duplex modes there should be a guaranteed echo loss of about 45dB (ITU-T Recommendation G.167). A new algorithm[10] for low cost fixed point DSPs allows to realize full duplex mode. The algorithm controls the echo canceller and loss control.

2.3 Where is the best place for a microphone?

The fundamental consideration during implementation of a microphone into an automobile is to find the best place. Sound intensity for a point source (a speaker is a point source) follows a 1/square law.

$$(1)\ I = \frac{W}{4\pi r^2}$$

Or in other words: doubling the distance between speaker and microphone will decrease the sound pressure level (SPL) by 6dB. As a consequence, signal to noise ratio or acoustical performance of the acoustic path will deteriorate by the given driving noise. An optimum has to be found for minimum distance: speaker to microphone and

- minimum distance speaker to microphone and
- maximum distance: noise sources to microphone

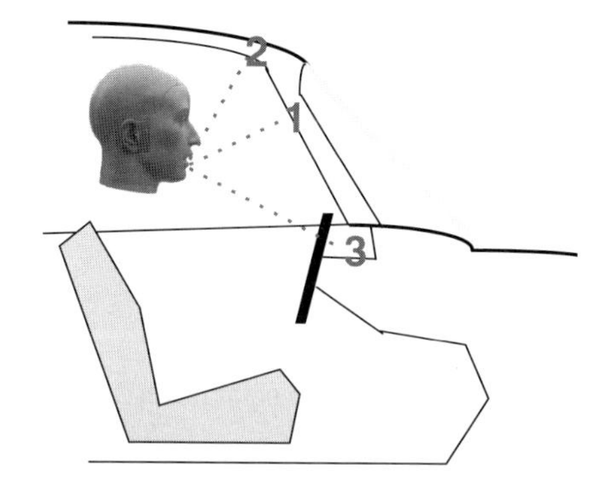

Fig. 2.10[2]: Possible microphone locations(side view)

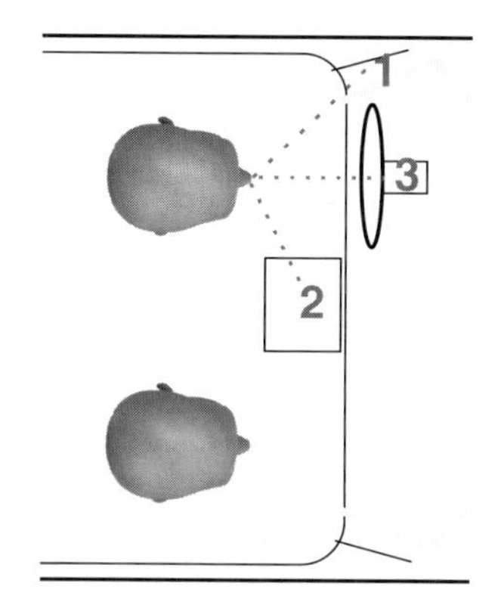

Fig. 2.11[2]: Possible microphone locations (top view)

Of course interiour styling and wiring harness will influence the possibilities. Every type of car and body design (sedan, truck, convertible etc.) needs a specific solution for the optimum microphone location. Variation in dimensions, distances, engines and noise sources between different kind of types change the system performance. Especially the interiour trim has a great influence. For example: Leather reflects sound waves to a greater extent in comparison to absorbing textiles.

Pos	Position	Distance Speaker/Mic	Noise Source
1	A-Pillar	51cm (-2.6dB SPL)	• body • wind • air condition
2	Headliner	44cm (ref. 0dB SPL)	• body • wind
3	Steering Column	48cm (-1.5db SPL)	• body • wind • steering column • defog • air condition

Table 2.1[2]: Possible microphone locations

2.4 Enhance speakers voice by microphone directivity

Once the best place is identified, the directivity of the microphone should be optimized. Omnidirectional microphones have a constant sensitivity over all angular directions. With that, they pick up all random noise - so their random energy efficiency is 1 (0dB).

In order to improve signal to noise ratio of the acoustic path, microphones should have maximum sensitivity in the direction to the speaker and minimum sensitivity to all noise sources. Low random energy efficiency provides distance gain at constant signal to noise ratio.

$$(2)\ distance\_factor = \sqrt{\frac{1}{random\_energy\_efficiency}}$$

(2) Distance factor in comparison to an omnidirectional polar pattern

Bidirectional microphones (Figure 2.12) could not be used in most locations because backwards and frontwards they have equal sensitivity. For example, the A-pillar position wind noise would not be supressed. Cardioids (Figure 2.13) may be used. In most cases hypercardioids (Figure 2.14) or supercardioids (Figure 2.15) are the best choice because they allow to supress loud single noise sources like defog or climate control openings by turning the position of the null into their direction. Sensitivity to the speaker remains quite reasonable by this strategy.

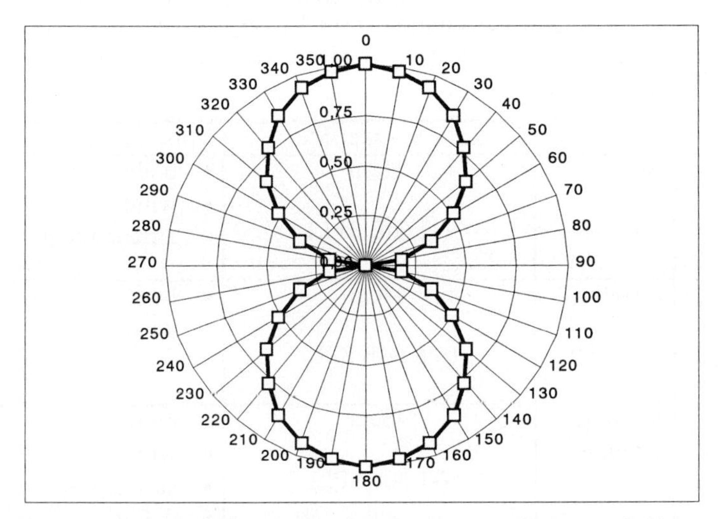

Fig. 2.12: Bidirectional directivity pattern (random energy efficiency is 0.333 or -4.8dB)

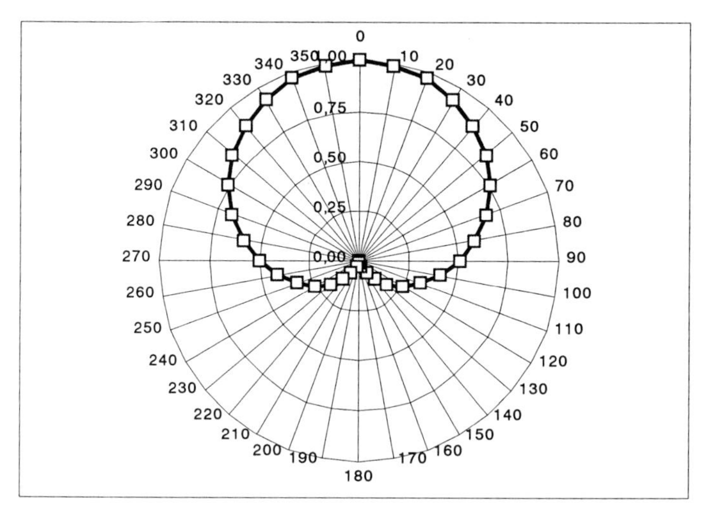

Fig. 2.13: Cardioid directivity pattern (random energy efficiency is 0.333 or -4.8dB)

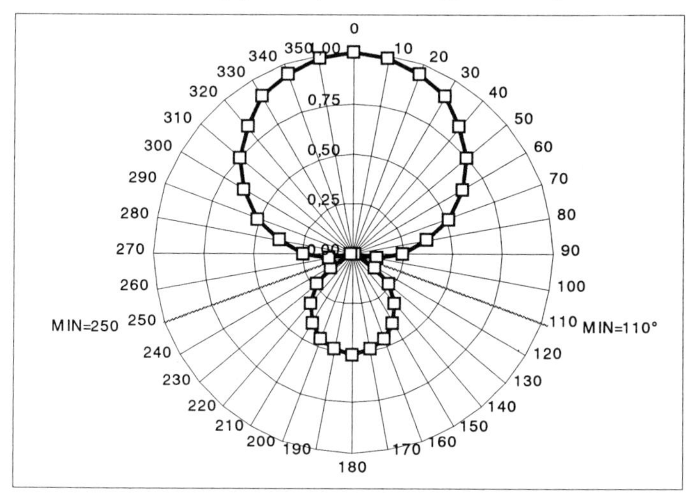

Fig. 2.14: Hypercardioid directivity pattern (random energy efficiency is 0.250 or -6.0dB)

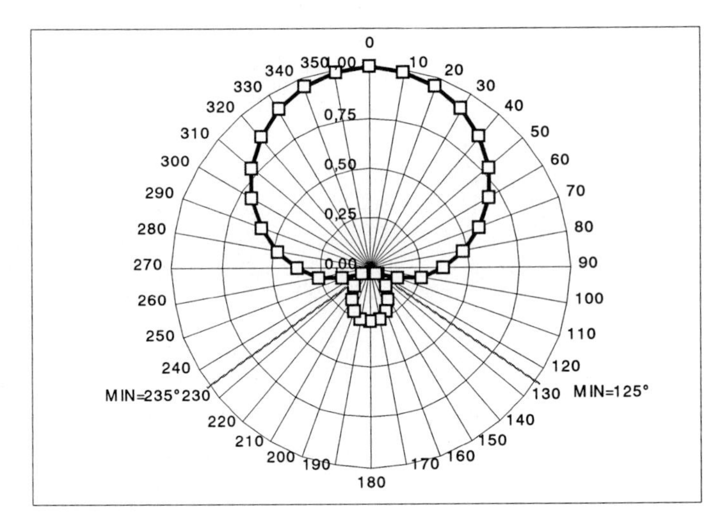

Fig.: 2.15: Supercardioid directivity pattern (random energy efficiency is 0.268 or -5.7dB)

For a microphone designer there are different methods to gain the desired directivity (Figure 2.16). The right choice is always a trade off between cost and performance. Advanced microphones (Chapter 3) will utilize especially multiple cartridge approaches with electronic directivity shaping.

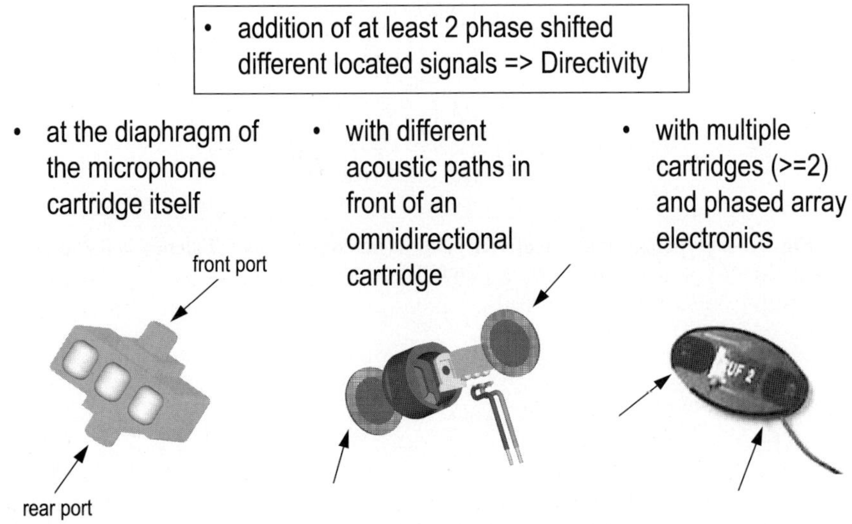

- addition of at least 2 phase shifted different located signals => Directivity

- at the diaphragm of the microphone cartridge itself
- with different acoustic paths in front of an omnidirectional cartridge
- with multiple cartridges (>=2) and phased array electronics

front port

rear port

Fig. 2.16: „Beam forming“ for automotive microphones

2.5 Frequency Response

Frequency response is important for many reasons and may have different demands according to the application:

- Cellular phones have limited bandwith from 300Hz to 3.4kHz. Active microphones should conduct active frequency shaping within these limits. Sensitivity curves should be balanced in a way to maintain speaker's sound color. The speaker will be recognized by his unfalsified sound color.
- Electronic gain and volume controls also need a bandwith limited signal due to reasons of digital signal processing and to make sure that noise sources outside of speakers' frequency range cannot be disturbing.
- Voice recognition and speaker identification systems need signals of very high quality because todays algorithms are behind the capabilities of humans.

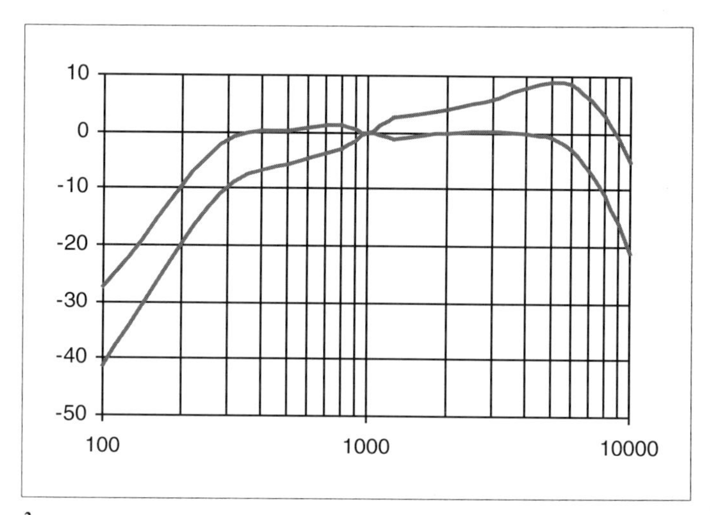

Fig. 2.17[2]: Frequency response slope channel of a state of the art active Ruf Electronics microphone, optimized for a luxury sedan for handsfree telephone application

2.6 Vibration Isolation of Body Sound

A car faces several vibration sources like the engine, electric motors and suspension movements. These vibrations are transmitted more or less into the body structure. Since handsfree microphones are mounted in locations within the structure it is evident that microphones should not be sensitive to vibrations, especially when they are in the audible range or at lower frequencies. This is for three reasons:

- The signal to noise ratio would be deteriorated
- Algorithms and electronic circuits which are connected to the microphone signal may be disturbed by unwanted pick up signals
- The microphone could be mechanically destroyed

Particulary when the microphone location is in the area of the dashboard, vibration isolation is very important. In the dashboard region the microphone may probably be mounted into plastic shrouds of the steering column or the dashboard itself. Because of the low distances to noise sources signal levels would be transfered at reasonable height. For example turning the steering wheel could transfer a scratch noise via the clockspring into one of the shrouds. Or the rotating fan motor would generate a sinus with higher order harmonics.

Shockmount microphones can be specifically designed in order to not exceed a 25mV AC limit while being shaken in each of the three room axes by 0.01g RMS over a frequency range from 50Hz to 1kHz.

2.7 Sensitivity to Wind - Air Stream Protection

Within a car there are two main sources of airstream which may disturb a microphone:

- Air stream caused by the convenient systems defog, defrost and climate control. The influence is expected to be great because the microphones are positioned in the windshield area. But there are chances to find an optimum between acoustic and convenient parameters.
- The other source is air stream which is generated by open window driving or by open tops in the case of convertibles. This situation is more difficult because the energy content of the noise is much higher especially at speeds above 50 km/h.

The effect of air stream which is picked up by a microphone is in generell a decreased signal to noise ratio and more serious a signal clipping which may occur in the electronics as well as at the diaphragm itself. This deteriorates speech intelligibility for recipients regardless whether humans or voice recognition systems. Test results at Ruf Electronics show, that there is a potential of up to 10 to 15 dB for improving airstream protection. Nevertheless voice pickup in a remarkable airstream remains a difficult task.

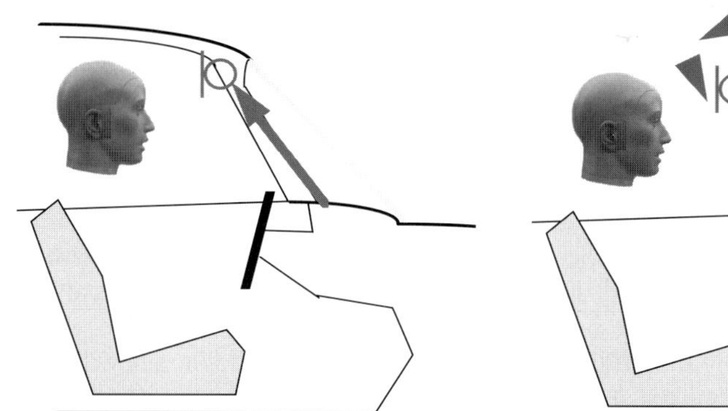

Fig. 2.18: Air stream caused by convenient system: defog, defrost, climate control ...

Fig. 2.19: Air stream caused by car design: convertible, open top, open window ...

Resumee

Many drivers made bad experience especially with some aftermarket handsfree kits[8]. There are some gaps regarding echo, noise and disturbances which may occur because the systems were not adopted to the special car model. This underlines the efforts for a design in solution.

Based on theoretical calculations driving tests at different speeds and road conditions have to be performed. The measurements of signal to noise ratio should be conducted in a way which allows to weigh the influence of comfort systems like defog and climate control.

Especially for handsfree communication applications a psychoacoustic test should round up the optimization procedure. This test makes people rate the understandability of recorded test speeches. These speeches were stored in different microphone locations and in a variety of driving and operating conditions.

3. Advanced Microphones for Handsfree Communication

Advanced microphones use multiple cartridges, which are switched together with a phased array electronics. The new generation of Ruf Electronics micRUFone™ (Figure 3.1 and third photograph of Figure 2.16) utilizes the differential signal of two omnidirectional microphones in a defined distance to each other. A time delay line in the path of one microphone produces a phase shift which results in a first order directivity pattern. This gives some very important benefits:

- The directivity is controllable by electronic means. This allows to implement a new microphone into a new car model in a shorter period of time. Directivity may be changed during the test with an electronic switch box. The conventional way to change directivity, as described in the first two cases of Figure 2.16, needs more complicated changes of the hardware.
- The microphone can easily be mounted behind the clothes of the interior trim. The passengers will not see the microphone. This is possible because both signals will be attenuated and delayed in the same way resulting in a stable directivity pattern. Conventional microphones will be influenced. This makes it more complicated to design a conventional hidden mount microphone.

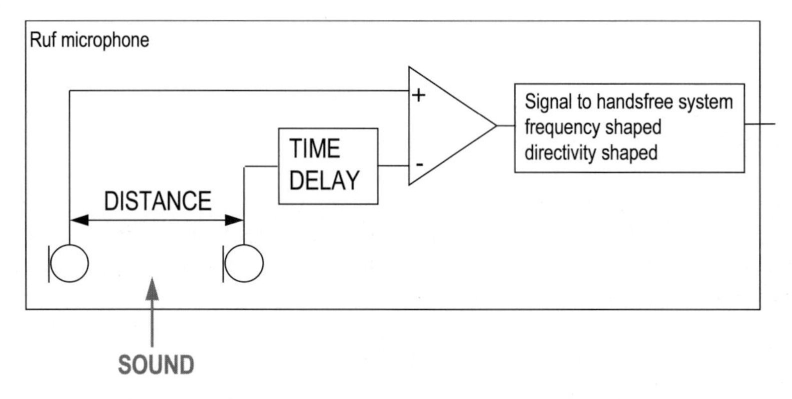

Fig. 3.2[2]: Simplified electronic to gain directivity within micRUFone™

Dual Output

It is obvious that such a layout gives space for additional improvements. A second signal could be generated which delivers the sound signal picked up with an omnidirectional characteristic. This signal measures the driving noise and may be used for the level control of the radio entertainment equipment.

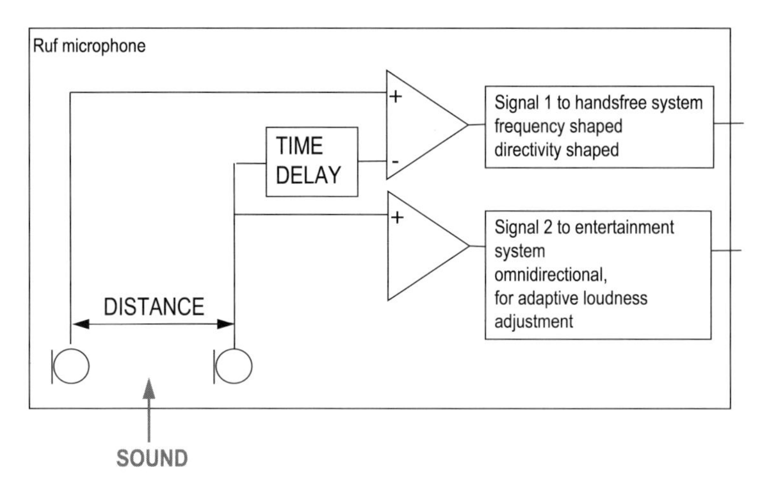

Fig. 3.4: Simplified electronic of micRUFone™, new generation microphone

Noise Cancellation

One of the most important topics for current research in the industry is noise cancellation. Digital Signal Processors (DSP) and capable algorithms allow one channel reduction approaches by spectral substraction. Several solutions show a noise reduction of up to 8 to 10 dB. This result can be obtained for constant and steady noise sources.

More realistic are mixed noise sources with steady content as well as with instationary behaviour like a passing truck or a speaking child in the backseat. The solution for this task will be a multi channel noise reduction system. Results from current dual channel systems demonstrate typical noise reduction figures between 10 to 12 dB. This depends strongly on the utilized algorithm and the current acoustic situation in the car. There is always a trade off between speech quality and noise reduction.

A typical microphone assembly of the near future will look like Figure 3.5. A DSP chip (> 4MIPS) uses an adaptive algorithm for directivity shaping as well as for noise reduction.

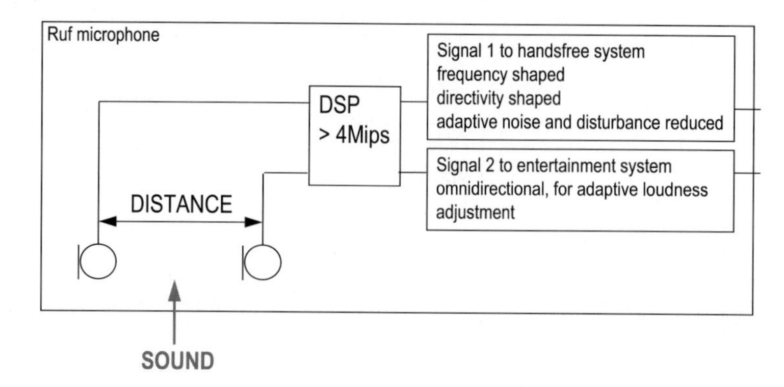

Fig. 3.5: Simplified electronic of micRUFone™, new generation microphone

Acknowledgement

The author would like to thank Robert Schulein and his team at ETYMOTIC RESEARCH for the efforts in a joint development program.

References

[1] V. Capel „Audio & HiFi Engineer´s Pocket Book“, Great Britain, 1995

[2] R. Frodl „New Generation of Active Microphones for Handsfree Communication and Voice Recognition“, SAE Paper, 1999

[3] NHTSA „An Investigation of the Safety Implications of Wireless Communications in Vehicles“, http://www.nhtsa.dot.gov/people/injury/research/wireless

[4] Ericson 1/98 - Figures End of 1997

[5] H.G. Burghoff et al „Ein Schritt ins nächste Jahrtausend - Elektrische und elektronische Innovationen“, Sonderausgabe von ATZ und MTZ: Mercedes-Benz S-Klasse, 1998

[6] G. Czymmeck et al „Die Zukunft hat schon begonnen: Kommunikations- und Informationssysteme“, Sonderausgabe von ATZ und MTZ: Mercedes-Benz S-Klasse, 1998

[7] VDA Verband der Automobilindustrie e.V. „Richtlinie GSM-Funktelefone, Funktelefonkomponenten und Funktelefonantennen“, Entwurf Dezember 1997

[8] V. Eklkofer „Handy-Einbau nach VDA-Richtlinie“, Funkschau 21/1998

[9] K. Jost „The car as a mobile-media platform“, Automotive Engineering International, May 1998

[10] T. Schertler, G. Schmidt „Implementation of a low-cost acoustic echo canceller“, Proceedings IWAENC-97, p. 49-52, London, 1997

Using in-Vehicle Systems and 5,8 GHz DSRC to Improve Driver Safety and Traffic Management

O. Clair
RENAULT - Direction de la Recherche
1 Avenue du golf, 78288 Guyancourt Cedex, France
Tel: 33 1 34 95 76 45 Fax: 33 1 34 95 77 22

Abstract

Renault, PSA, Cofiroute and CS route are developing in closed cooperation an in-vehicle information system for use on highways called AIDA. It is based on the standardized CEN-DSRC 5.8 Ghz read-write transponder technology. The system will be implemented on a section of the A10 highway in France (100 km between Paris and Orléans)

The proposed services intend to improve safety and to enhance drivers convenience. The system will furthermore assist the highway operator for road traffic monitoring and security management.

After a technical description of the system architecture and the on board terminals, this paper aims to present results of evaluation tests carried out since summer 1998. This includes figures on relevance and helpfulness of AIDA messages for drivers and subjective satisfaction and acceptance of AIDA.

In addition, an information is given on the European situation according to this type of application. The status of standardization, other similar projects and the European harmonization and development perspectives of such applications are described.

Keywords: Transponder Technology, DSRC, Traffic Management

Introduction

Currently, the program "AIDA" (Application à l'Information Des Autoroutes - Application to Motorway Information) is running within a consortium involving : car manufacturers RENAULT and PSA PEUGEOT-CITROËN, a motorway operator COFIROUTE and toll equipment manufacturer CS-ROUTE. The first project phase is funded by the French Ministry of Industry. This two years program started in September 96.

The objectives of AIDA project are to :

- Provide an on-board system (using 5.8 GHz Communications) capable of delivering real time information to drivers ;
- Provide information of traffic conditions (accidents, road works ahead), weather conditions (snow, rain, fog), and distress signal sent by crashed or damaged cars;
- Provide additional services such as distance to and identification of the next motorway exit, tourist and refueling information, estimated time of arrival and an indication of alternative routes ;
- Improve traffic data collection : speed, occupancy rate, incidents, stopped vehicles, weather conditions, etc., which will give to the highway operator a better knowledge of traffic conditions ;
- Produce a low cost terminal ;
- Launch a field trial, involving 500 equipped vehicles by on-board units, which aims to evaluate the overall integrated system.

The first developments based on this concept were performed in 1994-95 with the demonstration of the ADAMS (Automatic Debiting Application for new Motorway Services) prototype. The validity of the ADAMS concept has been demonstrated to 200 users, with the development of several traffic information services, and the implementation of DSRC beacons on a 15 km test site. The major strengths of the system are the automatic localisation of the vehicle as soon as it enters on the highway and the use of the vehicle as a "floating car". Thus, the bi-directional link between the roadside and the vehicle can be used to transmit from the vehicles to the road operator, information concerning road conditions : traffic, weather, incidents, etc.

System overview

The system architecture

The architecture of the AIDA system is made of an on-board system composed of a microwave transponder and a terminal, and a road communications infrastructure answering to the dual objective of performance and integration with the motorway system. The technical architecture of the AIDA system is shown on the figure 1.

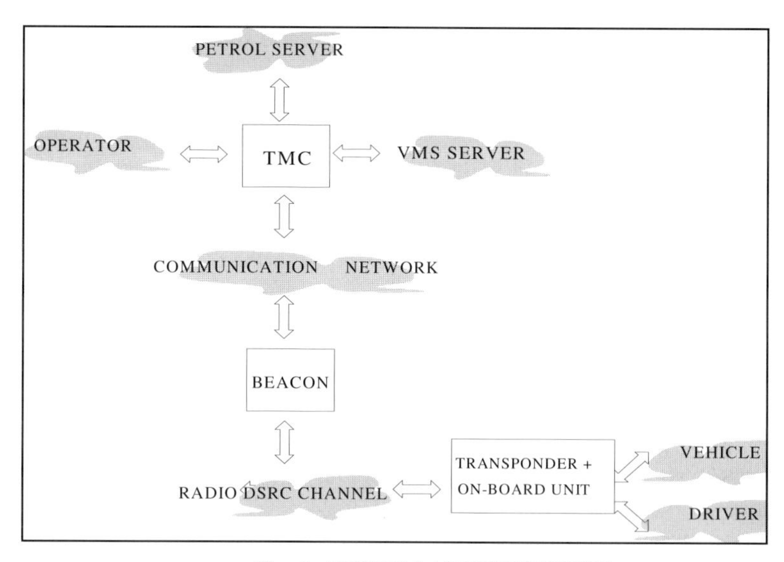

Fig. 1: SYSTEM ARCHITECTURE

Characteristics of the DSRC from an embedded point of view

The Dedicated Short Range Communication link (DSRC) is a short range multi-application radio link designed to exchange information between a transponder located inside the vehicle and a road-side beacon.

The information transmitted by a beacon about an event is described with a location expressed as a relative distance from this beacon. Information needs not to be immediately presented to the driver, as the timing for presentation to the driver is related to the nature of the information as well as the distance between the vehicle and the event.

This features is important, as it means that the place were the transmission takes place, i.e. the position of the beacon, needs not to be correlated with the nature and the location of the transmitted information. Thus, the driving factor for the installation of the beacons is mainly the refreshment rate of the information given to the driver than the exact position of the beacon.

Also, an advantage of such procedure is that the information is self sufficient. There is no need to have an on board database, or a particular knowledge, to decode geographic names. Drivers who are not familiar with the particular road environment have no difficulty to understand messages such as : « Road works 1 km ahead ». Information is given at the right time and there is no ambiguity on the location.

Another characteristic is that DSRC communication is a two way point-to-point communication. This enables to collect data from vehicles to the traffic control center. Highway operators get reliable, real time information on vehicle speeds, occupancy rates, weather conditions...

The AIDA services

There are three types of services on the AIDA system : safety, courtesy and operator services.

Safety messages stand out by immediate display on the on-board terminal and the transmission of an audible signal. This includes :

- information on traffic incidents
- information on traffic diversion
- information on weather conditions
- variable message signs
- exit signaling
- emergency call service.

Courtesy services may be consulted by the driver whenever he wishes such as :

- Exit distance and travel time information
- Price and petrol information
- Rest and service area information

Operator services concern data retrieval from vehicles through uplink communication

- Emergency assistance service to allow the driver to inform the traffic control center of an incident he encountered.
- Weather information collection.
- Travel time information collection.
- Slowing down information collection for automatic slow-down and incident detection.

The on board terminals

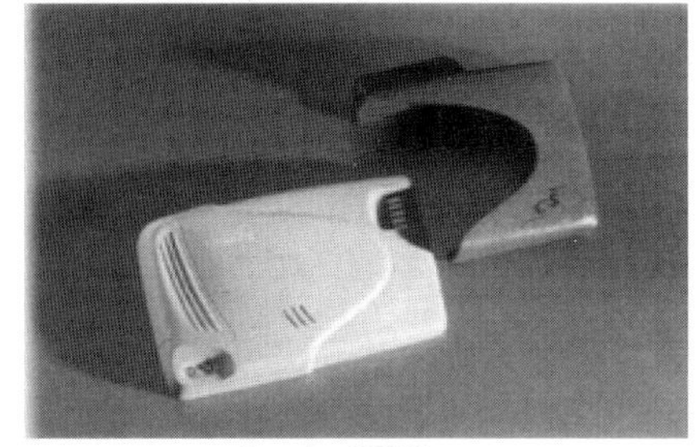

simple ETC tag

The first level of terminal is the simple ETC (Electronic Toll Collection) tags equipped with a buzzer or a LED. It supports only limited services to warn the driver in case of incident on the next few kilometers.

after sale terminal

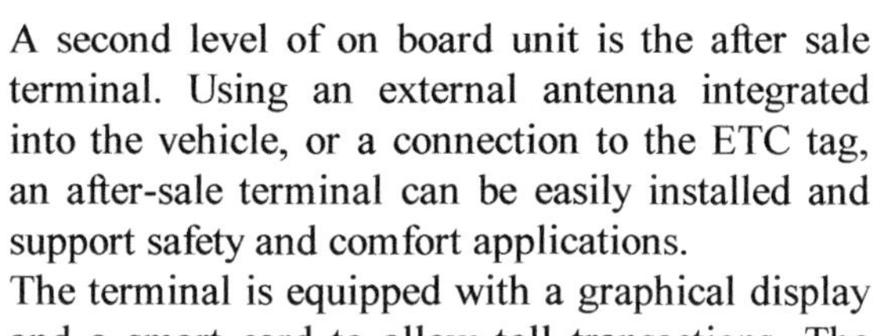

A second level of on board unit is the after sale terminal. Using an external antenna integrated into the vehicle, or a connection to the ETC tag, an after-sale terminal can be easily installed and support safety and comfort applications.
The terminal is equipped with a graphical display and a smart card to allow toll transactions. The interface with the vehicle is limited to power and speed sensor through the ISO interface of the car radio.

integrated terminal

The next step is the integrated terminal, available as an option in the vehicle. Full AIDA service is provided to the driver.
On the one hand, other applications such as navigation systems can benefit from this source of information. On the other hand a connection to other communication media like GSM and RDS-TMC, is possible to ensure service continuity from highways to suburban and urban areas.
And at last, as the on board unit is connected to different sensors in the vehicle (speed, ABS systems, fog lights, airbag ...) , reliable real-time information on motorway events is available to the road operator

First results of evaluation tests

Several evaluation procedures are carried out along the development of the AIDA system. A first one has been initiated in June 1998. It is characterized by an ergonomic evaluation. 40 subjects participated to these tests [1] . A second one, starting end 1999, will involve 500 regular users of the A10 Highway for the overall assessment of the AIDA concept.

From the first battery of tests, based on objective data such as driver's visual strategies as well as subjective data such as workload scale rating and usability questionnaires some preliminary conclusion can be derived.

Concerning the relevance and helpfulness of AIDA messages, users showed their satisfaction and interest. Furthermore, the system has demonstrated its usefulness for the subjects of the experiments while comparing behaviors of drivers (speed control, trajectory control, driving errors...) with and without AIDA for the same particular situation.
Preferred services are safety services.

For the man-machine interface, the first results let us to set up some future ergonomic recommendations in terms of message content and information presentation. This output is currently under construction.

This study has successfully prepared the future large scale evaluation of AIDA before launching the industrial phase. Further analysis will be conducted in order to assess impacts of the AIDA system use for drivers and the motorway operator.

Situation at the European level

Status of standardization

The major documents of the DSRC protocol have been voted in 1997. Communication layers are stable and the feasibility of concrete interpretability have already been proved.

On the system level, the core application, which is ETC, is well advanced. Several implementation projects are carried out in France, Norway, Netherlands etc. Concerning AIDA like services, the story is different. Harmonization projects across Europe are only starting now.

Other similar applications in Europe

Several countries have established test sites for Traffic and Traveler Information using 5.8 Ghz DSRC : France with the ADAMS-AIDA test site, UK where a new test site is to be established within Road Traffic Advisor (RTA) project, and Portugal where a test site is already installed in the area of Lisbon for LUSOPASSE system. In Sweden and Netherlands, test sites will be set to evaluate ISA (Intelligent Speed Adaptation) applications using 5,8 Ghz DSRC.

According to this situation, a European project called MARTA supported by DGVII on the TEN-T program, is starting in 1999. It should end by year 2001. The purpose of MARTA is to harmonize and standardize systems specifications and design, to ensure that these nationally developed systems are interoperable across Europe ; in addition, the systems must be capable of delivering information and route guidance in the driver's native language so as to achieve the potential benefits in the fields of safety, traffic efficiency and comfort for road users.

Conclusion

With the normalization of the 5,8 Ghz frequency for TTI applications, and the development of Electronic Toll Collection, DSRC has now become a very reliable communication link to provide the car driver with accurate real time information concerning his travel.

The AIDA project, and other European test sites, have shown the technical feasibility of such application and the interest for driver's safety and comfort. Taking advantage of development plans of ETC trough out the world, the next steps for the very near future concern harmonization and standardization work, with the MARTA project, and business development plans.

Acknowledgments

This paper described an experiment supported by the French Ministry of Transport - DSCR. I would like to thank all contributors to this paper and to the AIDA experiments. In particular G. Frémont and F. Belarbi from Cofiroute, JM Gautier from CS route, M. Vernet from INRETS/LESCO and Anne Ruthmann and M. Vassal from PSA Peugeot Citroën.

Reference

[1] « Traffic information with 5,8 Ghz DSRC systems : first results of the AIDA test trials. » G. Frémot, 5th World Congress on ITS - Séoul (Korea), 12 - 16 october 1998.

Intelligent Engine Control & Diagnostic Using the Ionization Current Sensing Technology

Joël Duhr
Forward Systems Development, Energy & Engine Management Systems
Delphi Automotive Systems
Avenue du Luxembourg
L-4940 Bascharage

Abstract: Instant feedback about the combustion process can be used for an optimization of the engine control and will improve the overall engine efficiency. During the combustion of an air/fuel mixture, ions and free electrons are generated. A current can be detected by applying a relatively low DC voltage to the spark plug or to another dedicated sensor. This ionization current contains information about the combustion process.
Through signal processing, information such as combustion quality, knock intensity, location of peak pressure, local A/F ratio, and others can be extracted. On a cycle by cycle basis this information is used as input for intelligent control as well as diagnostic, both for gasoline and diesel engines.
This technology used so far for engine and component development is now ready for mass production. Ionization current sensing has already demonstrated that it can easily survive the lifetime of the vehicle. Alternative technologies still have to prove themselves with respect to performance, robustness, reliability and cost efficiency.
In respect to the ignition functionality, to be able to satisfy future requirements on SI-engines, new technologies are investigated offering variable spark duration and spark energy.
Various system integration and partitioning alternatives are being considered. Developments in electronic circuit board technologies, as well as the evolution and availability of cost effective intelligent electronic components, do allow an integration of the signal processing part and other functions into engine mounted components, resulting in an optimization of an overall engine management system.

Keywords: Engine Control, Engine Management, Ionization Current

1. Introduction

The evolution with respect to emission regulations, system diagnostic requirements and fuel consumption reduction are requiring severe improvements in the area of engine control. To be able to fulfill those ambitious future goals, the overall engine efficiency has to be improved and a combustion process feedback seems to become indispensable. This information is then used as an input for optimized engine control strategies.

Delphi Automotive Systems started exploring the ionization current sensing technology, for engine control, about 10 years ago. [1] This technology used so far for engine and component development is now ready for engine control in mass production. Ionization current sensing has already demonstrated that it can easily survive the lifetime of a vehicle. Alternative technologies still have to prove themselves with respect to performance, robustness, reliability and cost-efficiency.

Beside gasoline SI-engines, this technology is also applicable to diesel, CNG, alternative fuels as well as 2 stroke engines. However, this discussion will concentrate mainly on gasoline spark ignition engines.

The ion-sensing technology is currently in low volume production with the following functionality:

- Base Functionality
  - Ignition
  - Misfire Detection
  - Cylinder Individual Knock Detection
  - Cylinder Identification

Delphi Automotive Systems has developed a sub-system offering these functionality's. Optimizations regarding secondary ignition energy, for SI gasoline engines, have been included, to satisfy future ignition requirements.
Furthermore, serious investigations are performed for additional engine control functionality's, using the ionization current signal. These investigations include optimization of the system partitioning and integration:

- Advance Functionality
  - Location of Peak Pressure for Ignition timing Control
  - EGR Control
  - A/F Control

2. Technology

In an ideal combustion, hydrocarbon molecules react with oxygen and generate carbon dioxide and water. During this process, there are also other reactions that go through several steps before they are completed. Ions are generated by the chemical reactions in the flame. [2] Additional ions are created when the temperature increases as the pressure rises and more ions are generated as the internal energy of gases increases.

The ions generated by the flame have different recombination rates. Some ions recombine very quickly to more stable molecules, while others have longer residual times. The most stable ions generate a signal that follows the cylinder pressure due to its effect on the molecule concentration. The high temperatures of the burned gases are also creating ions and free electrons.

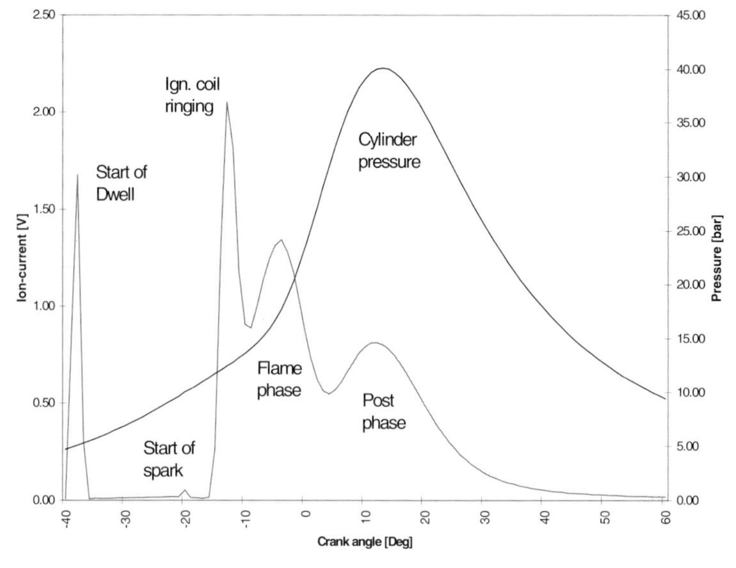

Fig. 2-1

Typically the ionization signal consists of 3 phases:

The first phase, ignition coil ringing, is caused by the dissipation of the remaining coil energy, in the ignition coil, at the end of the spark. It has no information content.

The second phase, flame phase, is the chemical ionization between the spark plug electrodes as a result of the burning of the air fuel mixture.

The third phase, post phase, or temperature ionization, is a result of the recombination of ions being created later during the combustion process. This recombination rate depends mainly on the temperature. Since the temperature follows the pressure during the compression of the burned gases, i.e., when the flame propagates outwards and the combustion completes, the resulting ionization current will hence depend on the pressure. Additionally, it was also found that a relatively minor species, NO, seems to be the major agent responsible for the conductivity of the hot gas in the spark plug gap. This is explained by its large difference in ionization potential in comparison to the other species in the combustion chamber. The NO formation rate is also much slower than the combustion rate and most of the NO formation takes place in the post-flame region.

2.1 Detection Circuit

The ionization current detection system uses the production spark plug both as an actuator, for combustion initiation and also as a cylinder individual intrusive sensor. The detection circuit has been conceived on the low voltage side of the secondary ignition circuit using the spark current to generate a required DC voltage, thus eliminating any high voltage components and a separate voltage supply. A relatively low bias voltage is applied to the electrodes of the spark plug. This bias voltage has to be of a certain level: High enough to create a detectable current flow, but low enough not to act as an electrostatic filter attracting soot from the combustion chamber resulting in a fouling of the spark plug.

The measurement on the low voltage side implies one single coil per spark plug. As the trend, in respect to ignition system requirements, develops, one coil per cylinder mounted directly on the spark plug will be the most effective approach.

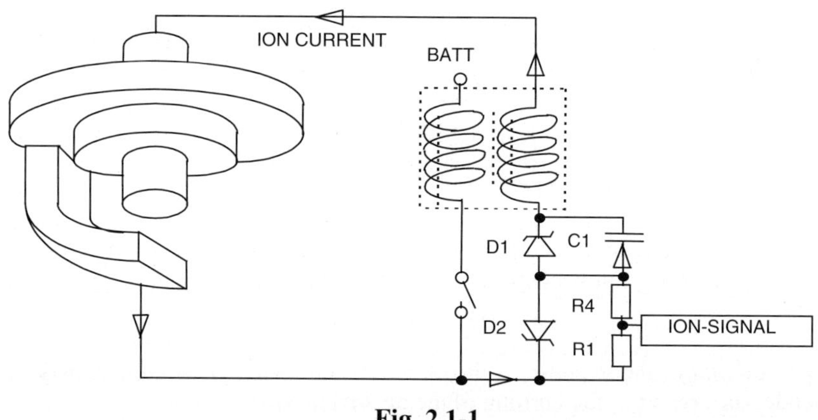

Fig. 2.1-1

In the ionization current sensing configuration, 1 intrusive sensor, the spark plug, is used per cylinder. It is a " direct " measurement correlating the ion current to the combustion process. However, the spark plug can not be used as an actuator and a sensor simultaneously, thus the spark duration has to be controlled not to loose any signal information, especially at high engine speed and/or low ignition timing advance.

3. Functionality

The ionization current detection system has multiple functionalities.

3.1 Ignition

One important functionality is the actuator, the ignition functionality. Three main ignition related parameters influence the combustion initiation and enhancement, i.e. the engine performance:

- Energy at the electrodes
- Electrode geometry
- Spark duration

In case of the ionization current detection technology, a trade-off between detection capability and spark duration has to be made as previously described. As the spark duration is mainly dependant of the ignition coil inductance, the ignition coil has to be designed for the worst case, i.e. maximum engine speed.

To guarantee optimum engine performance with reduced spark duration, other ignition related parameters may have to be optimized. The electrode geometry is the most appropriate to be adjusted. Increasing the energy by raising the secondary current is not recommendable because of the energy losses in the secondary resistance and it's impact on the spark plug electrode erosion.

It is known that a longer spark duration has a positive effect on the combustion process of a homogenous and/or a stratified diluted mixture. [3] This dilution being induced either by supplementary air or exhaust gas. As the engine is only operated at relatively low speed under those diluted operating conditions additional time is available to supply more energy into the combustion chamber. Delphi Automotive Systems is therefore developing a multi-charge, configurable spark duration, spark energy ignition system.

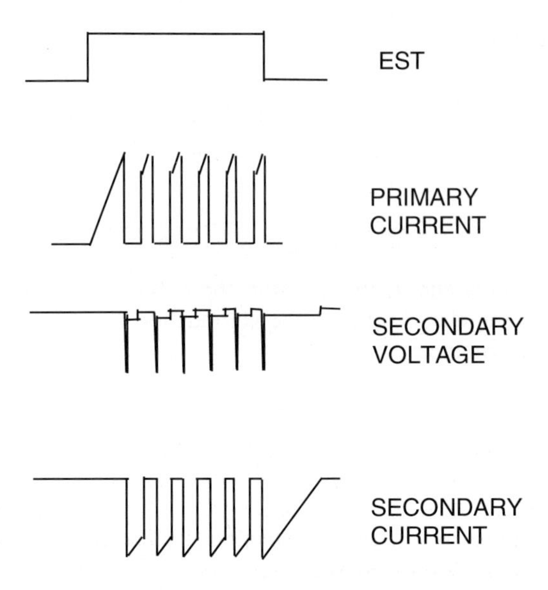

Fig. 3.1-1

In this case, an adapted and optimized ignition coil is charged and discharged multiply during one combustion process. This is a rather cost effective and efficient approach to achieve variable spark duration in comparison to an AC ignition or a spark "clipping". As the coil is not totally discharged at every time, a faster and higher energy level can be transferred.

3.2 Combustion/Misfire Detection

The first sensing functionality is the combustion detection. [4], [5] Using the crankshaft speed variation method for misfire detection is difficult under some engine operating conditions. Due to the inertia of the engine and the precision of the crankshaft signal, detection becomes very difficult, if not impossible, at high engine speeds and low engine loads. The difficulty increases with the number of cylinders and the vibrations induced externally.

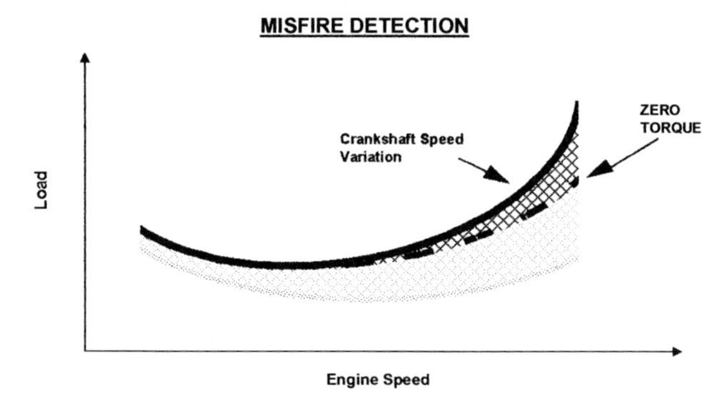

Fig. 3.2-1

The ionization current signal detecting the combustion is not influenced by external mechanical parameters. [6], [7] To detect misfire, no rough road sensing is required, as it is for a crankshaft speed variation detection configuration. The detection is also independent on the number of cylinders, the number of firing cycles per engine revolution and the inertia.

In case of combustion, ions and free electrons are created and can be detected. If misfire occurs, the ion current is zero. Compared to a pressure based system, which senses always the compression pressure, the signal to noise ratio, at low engine load, of the ionization current signal is superior. The detection range covers even negative engine load conditions.

As combustion events are detected the signal can also be used for synchronization of sequential fueling and ignition, replacing the CAM sensor functionality. The CAM sensor may be eliminated, as is the case in the current production application. However, a minimum number of firing events is required before the synchronization can be safely established.

3.3 Knock Detection

The knock information is filtered out of the raw ionization current signal. The combustion chamber knock center frequency for the bandpass filter is determined through a FFT of the raw ionization signal. No further application specific adaptations have to be done. In addition, no mechanical noises, generated through the valve train, engine wear, additional engine mounted devices, etc., are perturbing the detection. Therefore the knock detection system is absolutely high engine speed capable. The ion signal is also easier to process compared to a vibration-based sensor due to less dynamics in the signal.

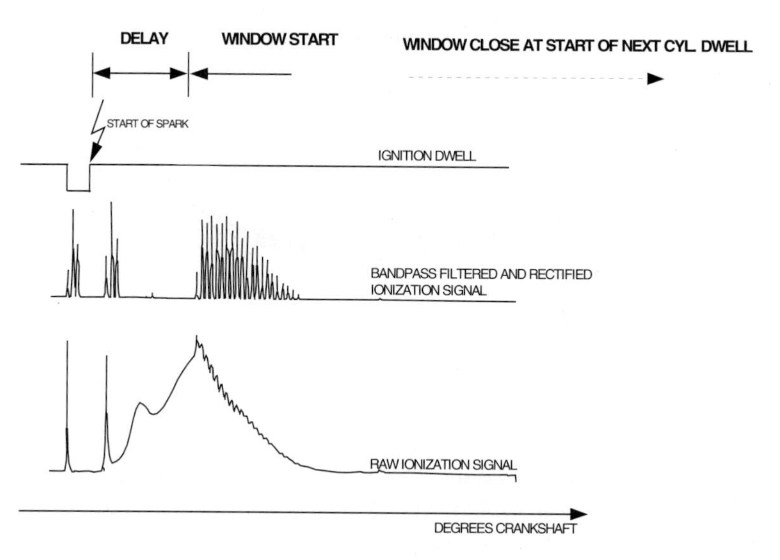

Fig. 3.3-1

A signal window strategy is applied to extract the knock intensity information. This windowing can be time-based and therefore simple to apply.

4. Advanced Functionality

The following advanced functionality's are currently under investigation by Delphi Automotive Systems.

4.1 Location of peak pressure

The location of peak pressure has a large impact on the engine efficiency (MBT). On today's engines these values are empirically defined and stored in a data table in the engine control unit. It would be ideal to have a constant feedback about the location of peak pressure to compensate for any variations (engine to engine variation, aging, deposits, intake air humidity, etc.). This would maintain the engine efficiency at the highest possible level.

As mentioned earlier, the pressure has most influence on the post-flame phase of the ionization current. Problems, however, occur when searching for the peak pressure position. A peak search is not really feasible since the flame-front phase often consists of more than one peak, and the post-flame phase often appears without a peak. Nevertheless, the ionization signal contains some information about the pressure, even if the post-flame does not contain a peak. [8] A model has been conceived to detect the location of peak pressure even under those difficult conditions.

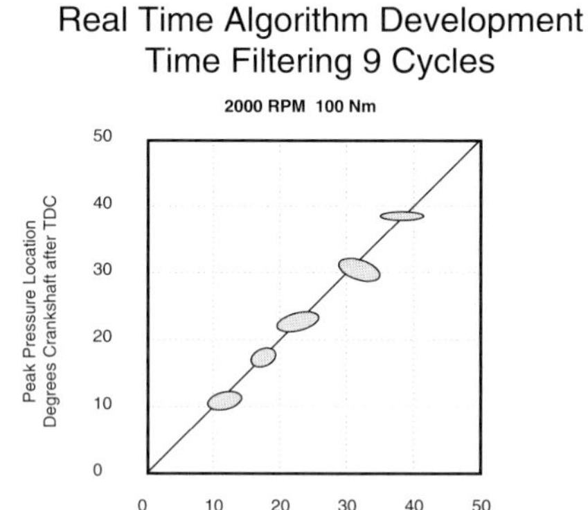

Fig. 4.1-1

So far, it has been demonstrated on an engine dynamometer to run the engine closed loop for location of peak pressure after TDC, using the ionization current as input. The best correlation to an intrusive pressure sensor was achieved by applying an 8 to 10 cycle running average filter. Around MBT timing, an agreement of 1 - 2 degrees crankshaft, between pressure based and ionization current based location of peak pressure has been achieved.

4.2 EGR / Lean limit

Similar to the ignition timing control, the exhaust gas recirculation rate is determined empirically and stored in the engine control module in a data table in most of today's control systems. No direct feedback, except maybe the EGR valve position, is available to compensate for the various system variations. Therefore, the control strategy has to be rather conservative and not all the benefits regarding fuel consumption and emission reductions are achieved.

It can be seen that the ionization current signal is strongly influenced by the recirculation rate of the exhaust gas. The slower burn rate at high EGR levels can be noticed in a delayed flame phase ionization peak, even with ignition timing and throttle compensated for the same location of peak pressure and engine load

Various properties of the ionization current signal were monitored and explored. Some correlate well to COV of IMEP, representing the driveability limit at high EGR rates.

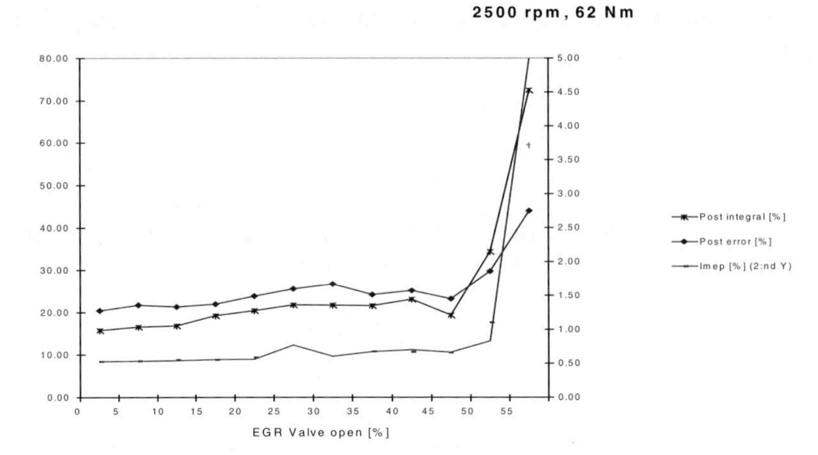

Fig. 4.2-1

Some other properties do correlate better to the exhaust gas recirculation rate.

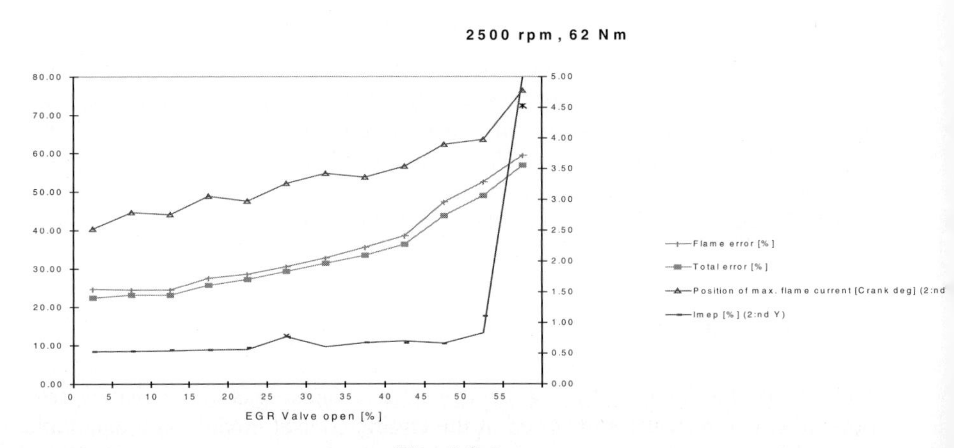

Fig. 4.2-2

To be noted, that the x-axis represents % EGR valve opening and not the absolute EGR concentration.

It can also be seen, that at least for this application and engine operating condition, a closed loop control via the pressure sensor would be unfeasible due to the non proportional increase of COV of IMEP with the exhaust gas recirculation rate.
Tests were made where the coefficient of variation in the ionization current integral was used as an input to closed loop control of the combustion quality. Using a variance controller, by first calculating the running average values and the deviation of the ionization signal as a normalized input to a PID regulator, allowed control of the EGR or lambda value to a certain degree of unstable combustion.

The ionization current integral has a qualitative connection to both EGR and lambda as well as IMEP.

4.3 A/F

Based on these findings, various properties of the ionization signal were analyzed to identify the parameter most influenced by the lambda value. This allows a calculation of the lambda value out of the ionization current.

Using the neural network technology, the maximum flame gradient has been defined as the best correlate to the lambda value. The maximum flame gradient stands for the largest positive slope on the filtered flame phase peak with a resolution of 1 degree crankshaft. With this information a stable closed loop lambda control from 0.8 to 1.3 could be demonstrated on a typical "lambda 1" engine.

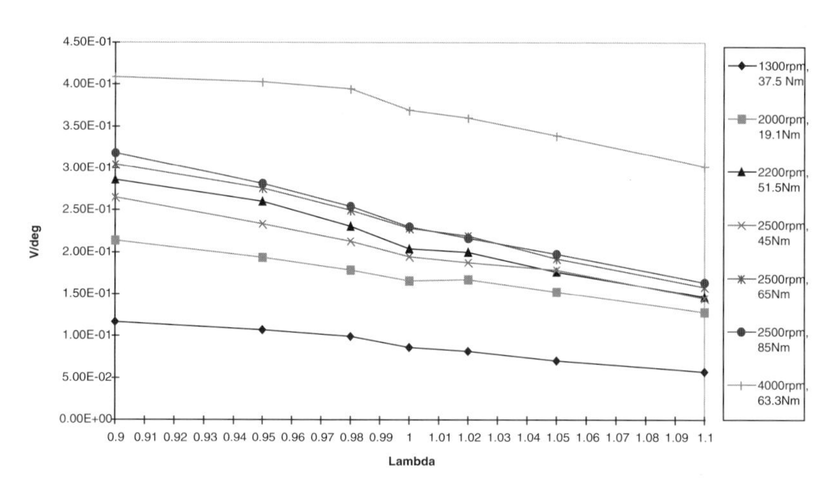

Fig. 4.4-1

Having one individual sensor per cylinder and an almost instant lambda information per combustion cycle, investigations were performed for an individual cylinder lambda control. This information was used to compensate for variations between cylinders. A control algorithm was developed to adjust the amount of fuel injected per cylinder to maintain the individual flame gradients at the same level. Compared to wide range engine oxygen sensors positioned close to the exhaust valve, a negative effect could be detected. With the controller enabled, the A/F variations between cylinder became larger. It is believed that this ionization current measurement is very local, i.e. in the spark plug gap. This local measurement may not represent the overall mixture in the combustion chamber. Further investigations are currently done to better understand this phenomena. It is believed that on engines with stratified mixtures, this local measurement could be an interesting function and feedback.

5. Implementation

The implementation and physical location of the ionization current sensing circuitry has to be done carefully, because of the low power level of the ionization current signal. In order to minimize any interference the sensing circuit should be as close to the ignition coil as possible, i.e. ideally underhood, engine mounted.

Various system integration alternatives are being considered, such as: Up-integration into an engine mounted engine control module, integration into an ignition cassette, as well as a standalone module. In the case of a sub-system configuration, the interfacing to the main controller can either be done through discrete lines or high speed serial links such as CAN.

Developments in electronic circuit board technologies allow high flexibility in respect to the system installation. [9], [10] Using a ceramic substrate of thick film, low temperature co-fired ceramic or high density technologies result in a cost effective, high temperature and mechanically resistant unit. Using this technology, the module has a very high packaging density and can be underhood or even engine mounted.

Technique	FR4	Thickfilm	LTCC	HD3
Size	100%	50%	30%	30%
Therm. Efficiency	-	+	o	+
Vibration Tolerance	o	+	+	+
Number of Layers	o	o	+	+

Fig. 5-1

The flip chip technology offers a tremendous gain in circuit board packaging density. In this case, the chip is turned around, "flipped" and soldered directly onto the substrate. The number of connections is reduced and thus a more robust and reliable design is guaranteed. Delphi Automotive Systems developed the process to be able to "flip chip" circuits even to the size of 32 bit microprocessors on a circuit board.

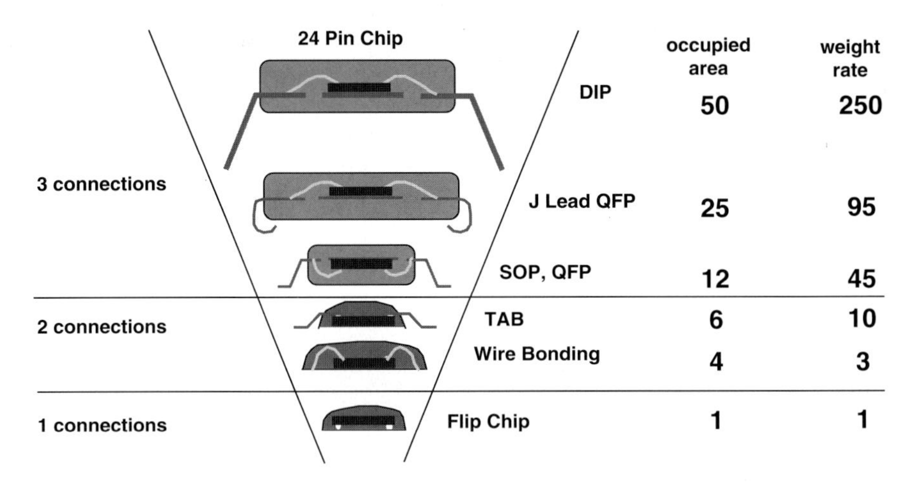

Fig. 5-2

Specifically for the ionization current sensing technology, Delphi has developed an integrated circuit, the IonIC, which performs the ionization current signal processing. This chip, based on the "flip-chip" technology, is designed to be controllable in terms of gains and filters allowing highest flexibility and covering a range of automotive and non-automotive applications. Furthermore, this circuit is designed to process a multiple number of cylinders.

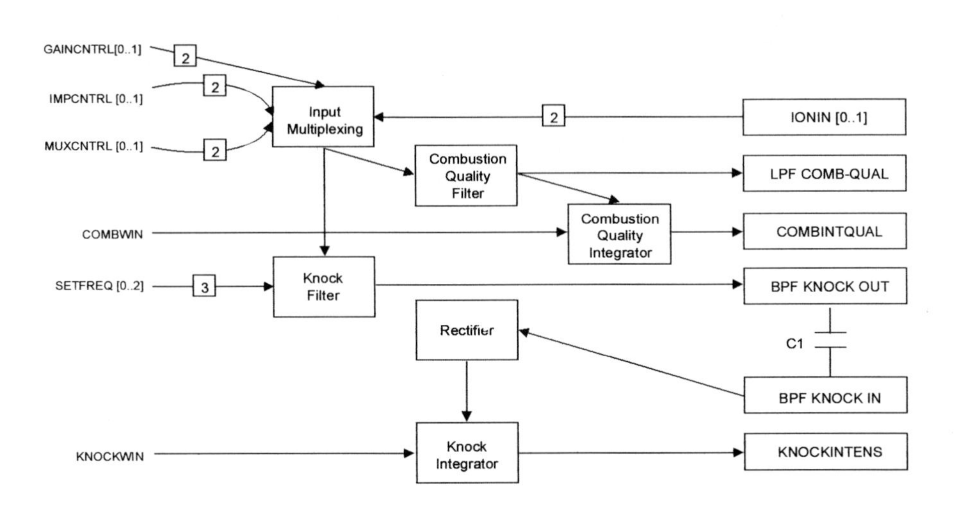

Fig. 5-3

The following picture shows a 6-cylinder ignition and ionization current sensing module. It includes the ignition coil drivers, the combustion quality and the knock intensity functionality. The interface to the engine control module is done via two PWM lines. One line transmits the combustion quality information and the second line transmits the knock intensity. To achieve the mechanical resistance and the packaging requirements this module is made of the "hybrid" technology, a ceramic substrate including flip-chip.

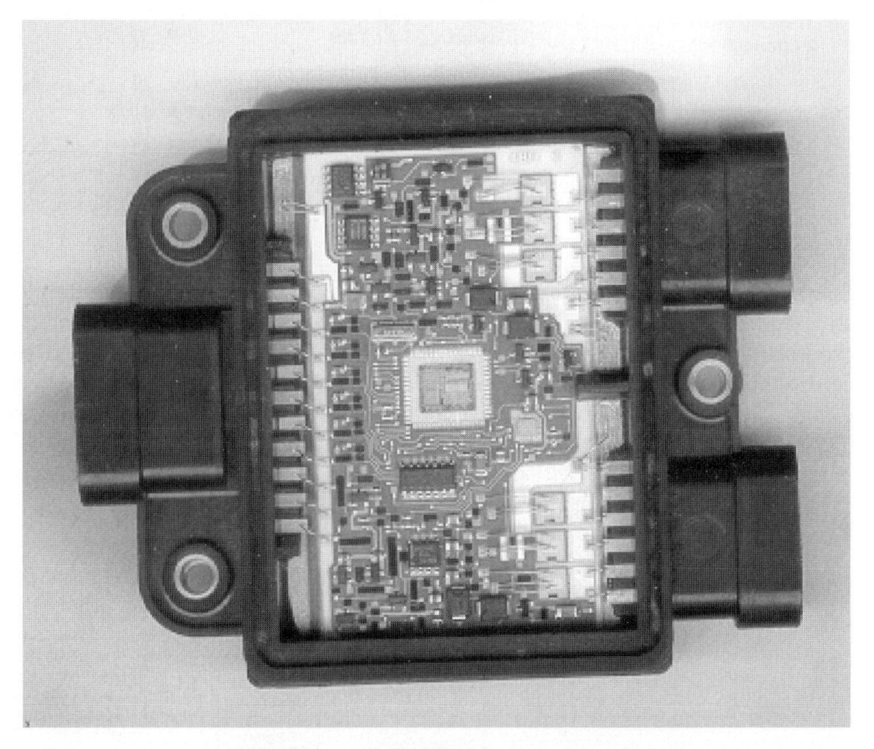

Fig. 5-3

With the evolution and availability of cost effective digital signal processing and neural network technologies, investigations are currently focused on exploring those capabilities in a mass production engine control unit. Using the ionization current as an input to a neural network with a simple single hidden layer configuration showed very encouraging results. It is believed that such a configuration could have a tremendous impact on the overall engine management system development time and performance.

Another interesting area for the ionization current technology is the " twin spark " combustion chamber design. In this configuration, even 2 sensors per combustion chamber are available to provide a signal feedback, allowing other measurements such as flame propagation in conjunction with an optimized and very flexible ignition source.

6. Conclusion

By using the spark plug as a sensor no additional components have to be installed in the combustion chamber. The sensing of ions and free electrons does provide a direct feedback of the combustion process. Currently in production, the ionization current system offers sensing of combustion quality, used for misfire detection and cylinder synchronization, and sensing of the knock intensity for cylinder individual knock control. In the very near future, additional functionality's such as location of peak pressure, EGR dilution, A/F information and ignition diagnostic will be available. With this capability, the overall engine control performance, robustness, as well as the development time, will be considerably improved.

References:

[1] Ion-Gap Sense in Misfire Detection:
Auzinsh J, Delco Electronics Co, Johansson H, Nytomt J, Mecel AB (February 27 - March 2 1995), SAE Technical Paper 950004

[2] An Ionization Equilibrium Analysis of the Spark Plug as Ionization Sensor:
Saitzkoff A, Reinmann R, Berglind T, Lund Institute of Technology, Glavmo M, Mecel AB (February 26-29 1996), SAE technical Paper 960337

[3] Internal Combustion Engine Fundamentals:
McGraw, Hill Series in Mechanical Engineering, Heywood JB (1988)

[4] Advanced Engine Misfire Detection for SI-Engines:
Föster J,Lohmann A, Mezger M, Ries-Müller K, Robert Bosch GMBH, SAE Technical Paper 970855

[5] OBD in a European context, 1st international Conference on Control and Diagnostics in Automotive Applications:
Hansen J, Delphi AS (Genova October 1996)

[6] Engine Misfire Detection by Ionization Current Detection:
Anson L (1995) SAE Technical Paper 950003

[7] Engine Control based on Ion Current Detection, The Automotive International Week, Seminars & Workshop:
Duhr J, Delphi AS (Torino, November 1996)

[8] Ignition Control by Ionization Current Interpretation:
Eriksson L, Nielsen L, Linköping Univ, Nytomt J, Mecel AB (February 26-29 1996), SAE Technical Paper 960045

[9] Electronics Integration and Architecture for Cost Savings and Increased Functionality:
Duncan C, Dr. Auzins J, Delco Electronics (1996), VDI Berichte Nr. 1287

[10] On-Engine Capable Engine Control Module - A Comparison Between Today's and Tomorrow's Technology:
Senninger H, Bibby R, Dr. Millen R, Delco Electronics (September 1996), VDI Electronics in the Vehicle

Micro-Injection System Using Ultrasonic Vibrations for Drop on Demand Ejection

Laurent Lévin , André Agneray
Technocentre Renault. Research Division / Electronic Dpt.
1 Av. du Golf. 78288 Guyancourt cedex. France.

Abstract: This micro-injection system consists of a *(50 mm long, 5mm wide, 570 µm thick)* silicon beam comprising a micro-structure with a glass cover plate on one side and a piezoelectric plate bonded on the other side. The micro-structure is composed of a feeder canal transforming right across in several parallel channels which end individually at the extremity of the beam thus forming a multi-orifice line.
An electrical signal is applied to the piezoelectric plate at a frequency which generates a flexural mode of vibration in the beam. The vibrations are mechanically amplified in the beam part comprising the micro-channels to give rise to large amplitude oscillations at the extremity of the beam where the micro-orifices are located.
The gasoline is delivered in the feeder canal and is naturally supplying the micro-channels thanks to capillarity and wettability. The liquid is permanently present in the vicinity of the micro-orifices and each oscillation of the beam generates the ejection of two drops for each orifice. The diameter of the drop is calibrated by the size of the orifice and the amplitude of the oscillation.
Thanks to a short transient response (10 to 20 µs), this type of multi-point fuel injection system can allow real time modulation of the amount of fuel injected during the admission cycle. It also generates a spray of micro-drops of a predetermined size that can be as small as 10 *µm*.
Those two essential caracteristics can greatly enhance the fuel preparation process which has a direct impact on the combustion process and consequently on the fuel consumption and exhaust emissions.

Keywords: Droplets, Microejection, Drops on Demand

1. Introduction

New designs in automotive engines are concerned with the decreasing of fuel consumption and the reduction of emissions. The developments thus integrate considerations for a higher efficiency of the combustion process which is actually a vapor phase process. Atomization plays an important role as the evaporation rate can be greatly enhanced by reducing the droplets diameters. Improved sprays reduce fuel-air mixing time, impingement on metallic surfaces and allows more uniform fuel-air mixture. Fast response of the ejection system also facilitates the capability to operate stratification [1].
The developments in spray atomization technologies usually involve hydrodynamic and aerodynamic perturbations to eject drops and ligaments.
In this paper we present a different method to generate drops with a piezoelectrically driven micro-actuator, the size of the drops being controlled by the driving frequency of the actuators and the amplitude of vibrations of the nozzles. This device compared to further developments [2] can lead to high fluid rates.

2. Device Schematic

Figure 1 is presenting a section of the overall device. The silicon beam is 50 mm long, 5 mm wide and is connected to the liquid supply at the rear part. The thickness of the beam is 570 μm except for the portion comprised in between the ejection nozzle and the piezoelectric ceramic where the thickness is varying from 100μm to 570 μm (length 10 mm) thus forming a mechanical amplifier. The feeding part takes place from the rear part to the beginning of the amplifying section and is composed of three canals (1 mm wide, 200 μm deep) borded with four beams (500 μm wide, full thickness 570 μm) supporting a glass cover plate closing the structure.
Those canals are evolving continuously (cf figure 2) into twenty micro-channels having a V groove section varying from 200 μm wide and 140 μm deep to 50 μm wide and 35 μm deep at the end of the amplifying section (10 mm long). The structure is closed by anodic bonding of a Corning glass (100 μm in thickness) and the terminations of the channels are forming twenty micro-orifices having a triangular section of the same geometry (50 μm wide, 35 μm deep).
A PZT plate (15 mm long, 5 mm wide, 0.8 mm thick) is bonded on the silicon and drives the structure in a flexural mode of vibrations. The mechanical oscillations are amplified at each step of the section containing the micro-channels and are maximum at the nozzles.

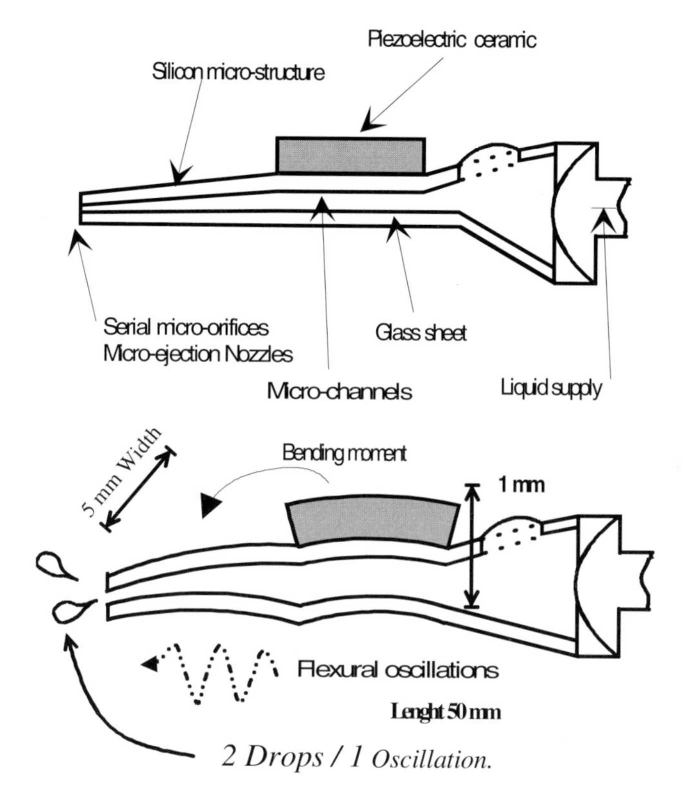

Fig. 1: Schematic of the device composed of the silicon and the piezoelectric plate.

3. Principle of Ejection

The micro-structure is full of liquid that can be supplied with overpressure if added valves are integrated in the structure to regulate the flow getting in. In our case, the liquid is permanently present at the back of the beam with a supply pressure slightly the same as the pressure of the external environment in which the device is ejecting.

Liquids like gasoline have the ability to wet almost any surfaces so the micro-channels get directly full of liquid by capillary effect once that kind of liquid is present at the rear part. Clear cuts at the end of the channels let sharp edges around the nozzles where the liquid is then stopped and forms a meniscus. When the nozzle is oscillating, the acceleration is inducing the extraction of a small amount of fluid from the column

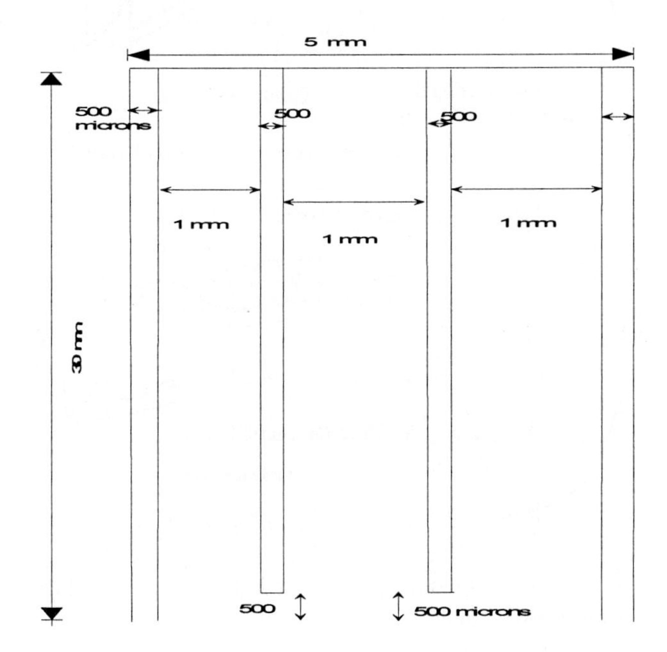

Fig. 2: Etching of the silicon beam. Feeding part

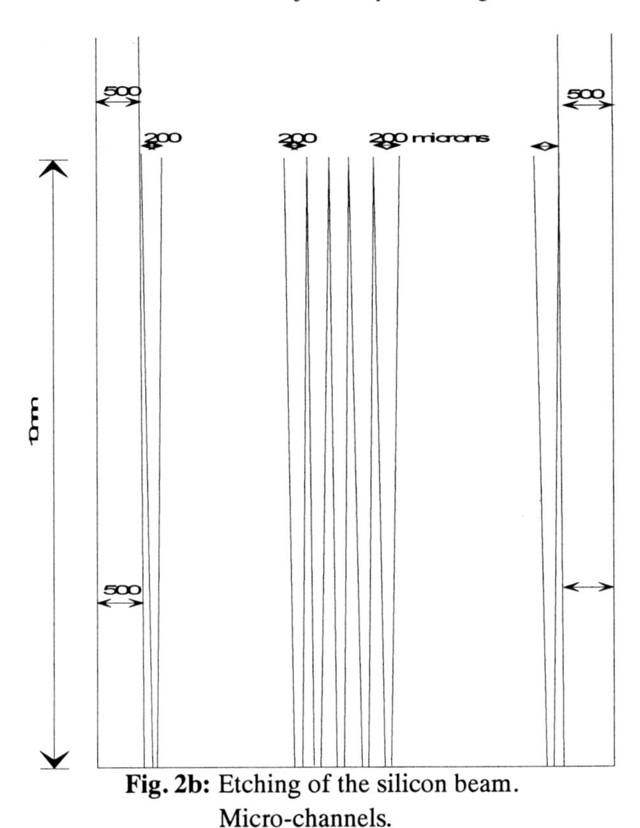

Fig. 2b: Etching of the silicon beam.
Micro-channels.

Fig. 3 is showing a sequence of 22 μm diameter drops ejected from a nozzle at a speed near 10 m/s in a medium regime where one drop is generated.

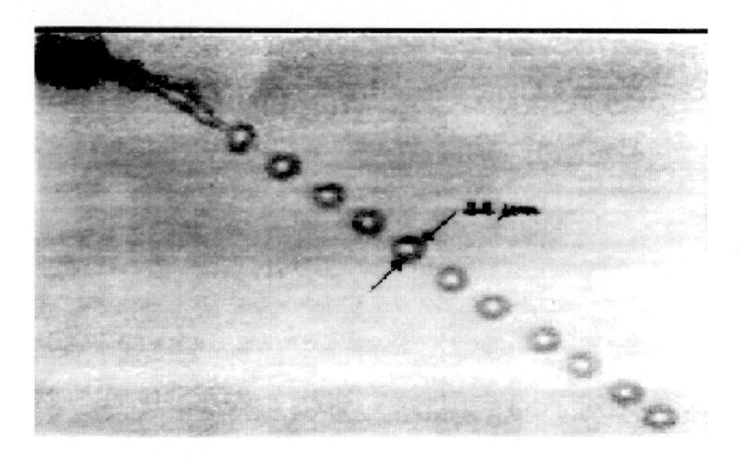

Fig. 3: Ejection of drops at a speed of 10 m/s in a medium regime. Diameter of the drops: 22 µm.

4. Dimensions and FEM Analysis

The transducer is designed to get maximum displacement at the end of the beam at frequencies higher than 100 kHz.
First dimensions have been carried out with analytical calculus to evaluate the wavelengths in silicon beam.

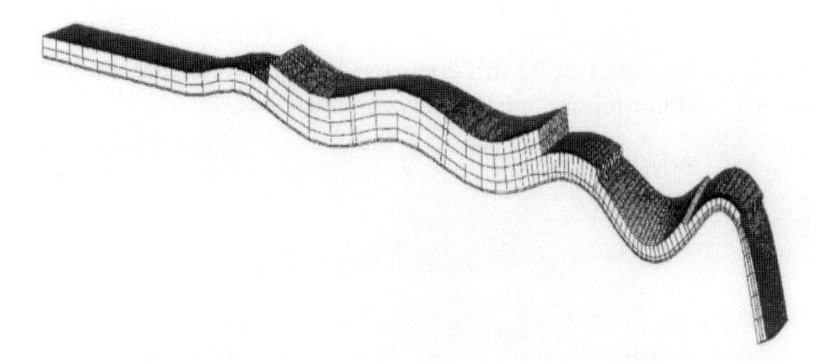

Fig. 4: Silicon micro-structure with the piezoelectric plate bonded and three amplifying steps oscillating at 116 kHz in a flexural mode.

The amplifying part is composed of three steps representing each one half-wavelength for the corresponding thickness, the last step being a quarter wavelength.
The calculation of classical dispersion curves for Lamb waves leads to the phase velocity versus frequency and thickness of the plate. Theoretical and experimental results are in good agreement, the grooves in the beam having a slight effect that is not taken into account in the calculus.
Typical values for the wavelength of that first antisymetric mode are as examples 5.5 mm at 130 kHz for 370 μm in thickness and 3.7 mm at 200 kHz for 200 μm
Displacements achievable with different size of PZT plates under semi-static regime for a given electric field has been also calculated.[3,4]. The bimorph part of the transducer can generate a displacement of 4 μm under 120 Vcc,. This displacement is then amplified through the diminution of the thickness of the beam and the resonance of the structure. Silicon has a very high quality factor thanks to a very low density of internal dislocations. Before failure it can be driven in amplitude eight times higher than Titanium alloys for example [5].
Also those calculus are helpful to determine the first dimensions, the complexity of the structure required finite element analysis to determine the resonant frequencies and adapt the geometry of the piezoelectric plate to optimize the displacements.

We used TECHSOFEM [6], a commercial code from TECHSONIC , to simulate the device with all microscopic details of the structure.
Fig 5 is showing a few resonant frequencies between 100 and 200 kHz with the corresponding values of displacements at the nozzles for 120 Vcc applied on a 0.8 mm PZT plate. The modes numbered 1, 4, 5 are flexural modes, the other ones presenting a coupling with transversal modes inducing some torsional modes.

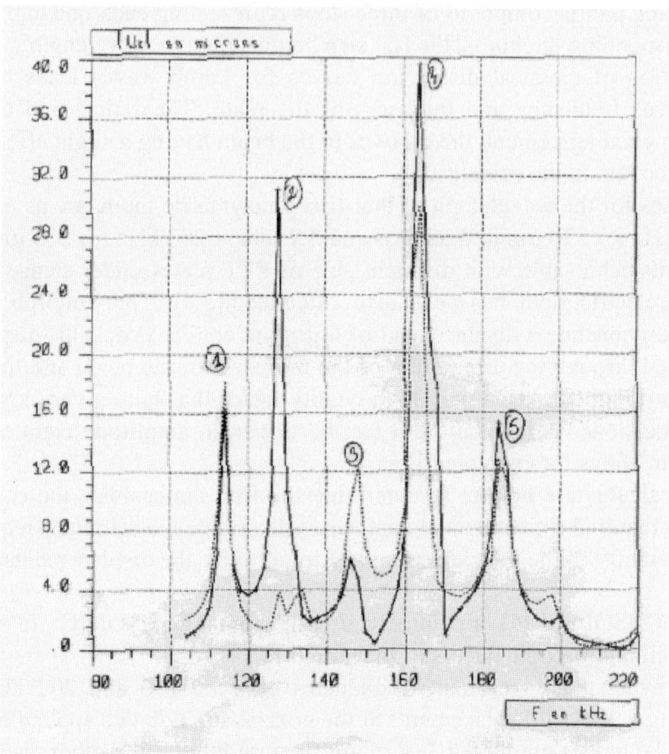

Fig 5: FEM analysis, resonant frequencies of the structure and the corresponding displacements at the nozzle location

5. Device Fabrication and Experiments

The silicon beam microstructure and the steps have been micromachined using KOH etching and several masks. Anodic bonding of a Corning sheet of glass is closing the structure. Final devices have been polished at the nozzle location to get sharp edges and adjust alignments of silicon and glass. A PZT 5H plate is then bonded with strong silver epoxy on the silicon and fine soldering contacts are operated. The micromachining has been performed with the CETEHOR [7].
Experiments have been carried out around 120 kHz leading to medium range ejection with 60 Vc driving. Higher excitation is leading easily to failures at the final step due

to either internal residual stresses during anodic bonding or micro-dislocations appearing during the cutting process. The amplifying part can be machined in the shape of a two dimensional horn type instead of reducing the thickness, allowing stronger devices for high fluid rates.

6. Conclusion

A new type of microejector has been developed that can be silicon micromachined. We designed the device to oscillate in a flexural mode and to have maximum displacement at the end of the beam so as to generate drop ejection . This type of system can be used to eject and pump many liquids with substantial fluid rates depending essentially on the degree of integration of micro-channels.

References

[1] "The spray characteristics of automotive port fuel injection - A critical review". Fu-Quan Zhao , Ming-Chia Lai and David L.Harrington. SAE Technical paper series.950506

[2] "Piezoelectrically actuated transducer and droplet ejector". Gökhan Perçin, Laurent Lévin and Butrus T. Khuri-Yakub.
Proceedings of 1996 IEEE International Ultrasonics Symposium
3-6 Nov.96, San Antonio, Texas.

[3] "The constituent equations of piezoelectric heterogeneous bimorphs" Jan G.Smits and Wai-Shing Choi. IEEE Transactions Vol.38 N°3. May 91

[4] "The constituent equations of piezoelectric bimorph"
Jan G. Smits, Susan I.Dalke. Sensors and actuators A, 28 (1991)

[5] "Micromachined silicon ultrasonic atomizer"
Amit Lal and Richard M.White

Integrated Engine-CVT Control

Mikio Nozaki and Masaru Mizuguchi
Nissan Motor Co.LTD.,6-1, Daikoku-cho. Tsurumi-ku, Yokohama, Japan

Abstract

Lately, improvement request of fuel economy gradually increase with the earthly environmental problem, while improvement of acceleration performance is always desired . Therefore, we developed "Integrated Engine-Transmission Control System" for compatibility of those requests. Combination of engine torque control with "Electrically Controlled Throttle" and CVT enables fine powertrain control based on the demanded drive torque. Good fuel economy can be compatible with driveability through integrated engine-CVT control, because these two performances can be designed and tuned independently. Moreover, the dynamic characteristics of driving force response can be improved.

Keywords: Engine Control, CVT, Fuel Economy, Powertrain

Introduction

Driving force of automobile is determined by the product of engine torque and transmission gear ratio. Combination of engine torque and gear ratio is limited with conventional powertrain unit, while fine and optimum control has been realized with the unit which has ways of torque control like Electronic Throttle and has CVT.

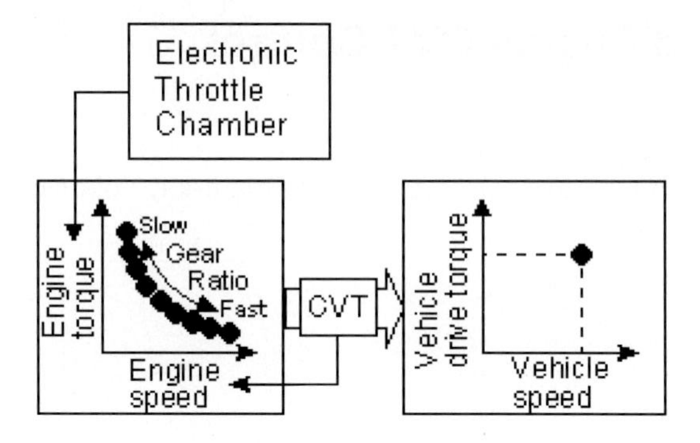

Fig. 1: Combination of ETC with a CVT

As shown in Fig.1, the drive torque demanded can be realized by choosing a driving point on the hyperbola curve ,on which the product of engine torque and CVT ratio is constant. Both engine torque and CVT ratio can be controlled continuously, so that we can choose minimum fuel consumption drive point at all times relative to the various levels of drive torque demanded if they are controlled in an integrated manner.
In addition to the cruise control system and the traction control system, more intelligent and complicated drive control systems like automatic drive or vehicle dynamic control are expected to grow rapidly. In those cases, it's good enough to adopt integrated engine-CVT control based on demanded drive torque, as the control system can be arranged simply and each systems outside can be connected or disconnected easily. Fig.2 shows a conventional vehicle/powertrain system. Fig.3 shows a future vehicle/powertrain system.

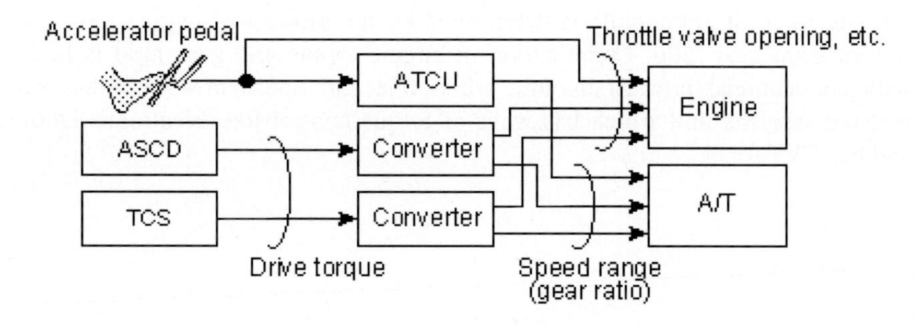

Fig. 2: Configuration of a conventional vehicle/powertrain control system

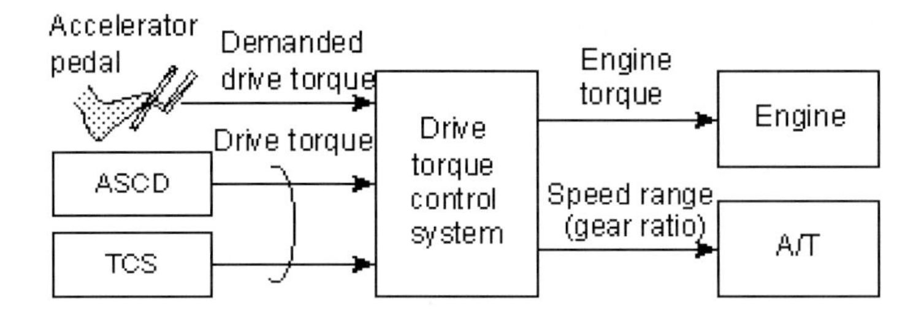

Fig. 3: Configuration of a future vehicle/powertrain control system

This paper presents an integrated engine-CVT control system that has been developed to obtain the demanded drive torque in the most efficient manner.

Control System Configuration

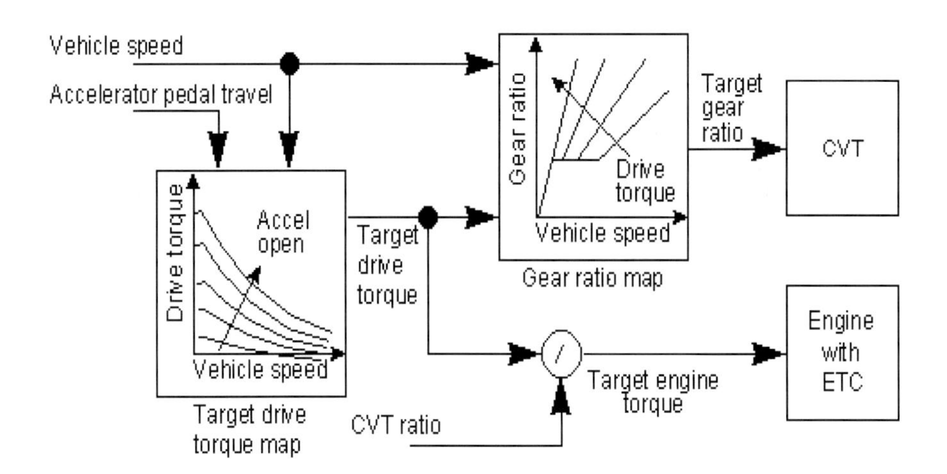

Fig. 4: Control system configuration

Fig.4 shows the outline of control system configuration developed. Target drive torque is calculated with accelerator pedal travel and vehicle speed. Target gear ratio is calculated with target drive torque and vehicle speed and is used for feedback control of CVT gear ratio. Target engine torque is calculated with target drive torque divided by real gear ratio which is obtained by CVT output and input revolution sensors. The target gear ratio is given by a map, which can be arranged for the fuel consumption to be minimum. An minimum fuel economy curve can be drawn on the map of steady state characteristic as a result of trace on the ridge of fuel economy.(Fig.5)

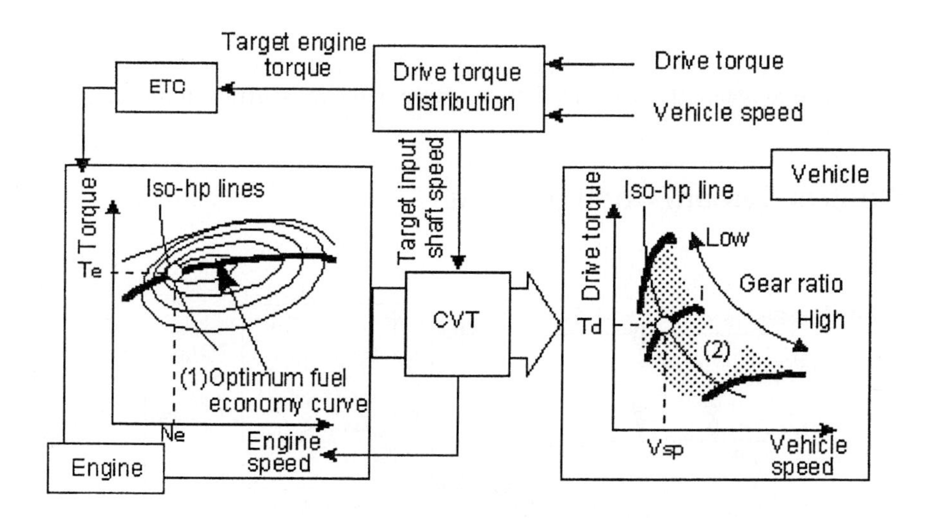

Fig. 5: Operation for optimum fuel economy

With conventional systems, characteristics of acceleration and fuel economy have large trade-offs. The integrated engine and CVT control system we have developped makes it possible to improve fuel economy by applying the gear ratio map, not influenced by the tuning of drive torque characteristics.

Merits of the System

Following three can be listed as merits of using this algorithm.

1. Compatibility of good fuel consumption rate and drive torque design.
2. Improvement of transient drive torque response.
3. Arrangement of communications for advanced vehicle drive control in the future.

Compatibility of Good Fuel Consumption Rate and Drive Torque Design

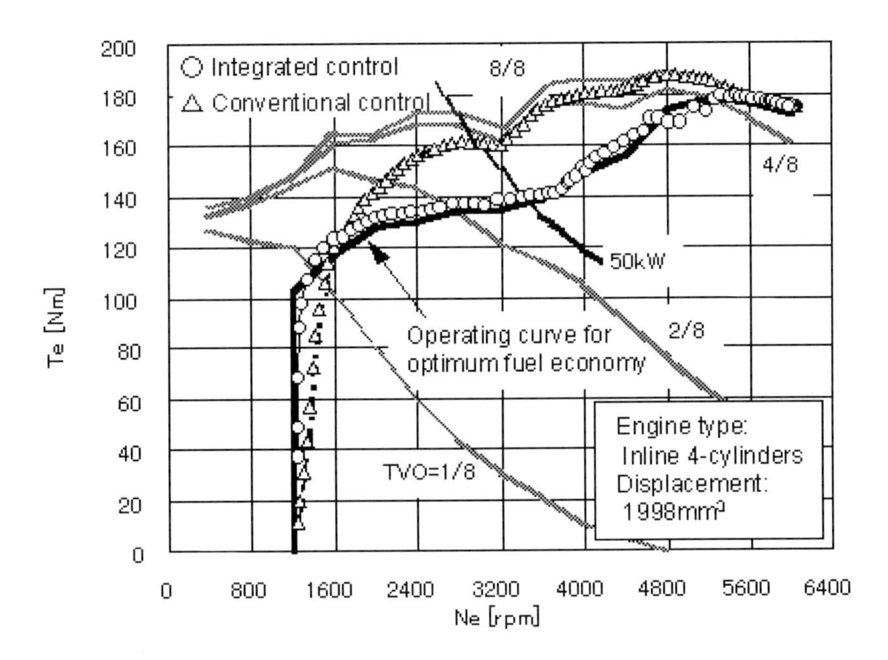

Fig. 6: Operation for optimum fuel economy

Fig.6 shows the comparison of driving trace lines between this algorithm tuned for best fuel economy and conservative system. When integrated engine/CVT control was performed (open circles), the data nearly trace the optimum fuel economy curve as indicated by the boldface solid lines. Even with such large accelerator pedal travel as the engine operating point falls in a power region where fuel economy would deteriorate with conventional system , it can avoid tracing such region with integrated engine-CVT control , so that the fuel economy can be improved. As a specific example, the integrated control algorithm reduces fuel consumption by approximately 8% when the required power output is 50kW.

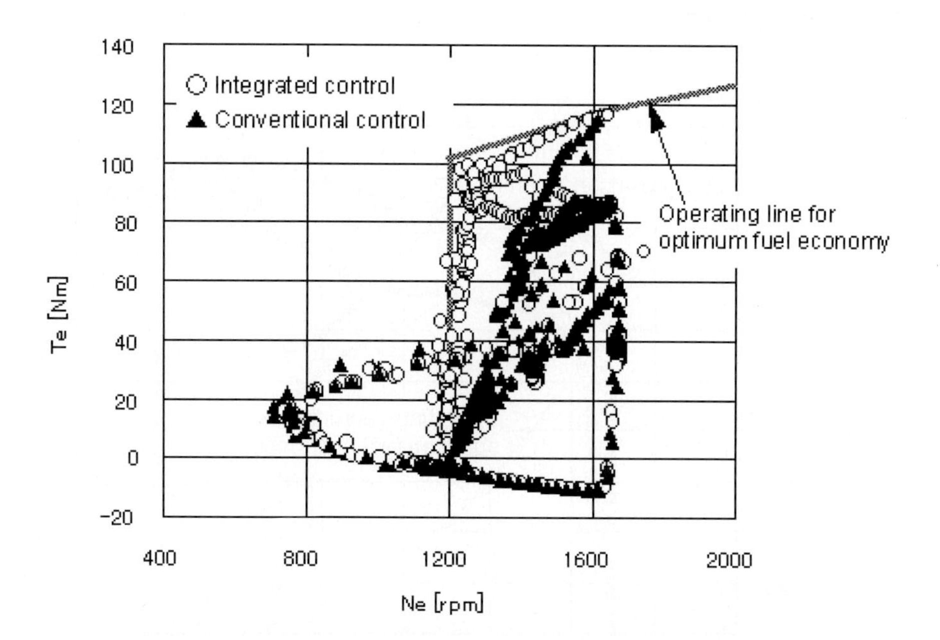

Fig. 7: Engine operating point under 10-15 test mode

Fig.7 shows the operating trace line under Japanese 10 and 15 test modes, which are the representative test procedures used in Japan to evaluate fuel economy as a result of simulation. This result indicates that engine can operate nearer to the best fuel economy trace line with integrated engine-CVT control system than with conventional system. As a result, effect of fuel economy by this algorithm is estimated by approximately 1.7%.(Fig.8)

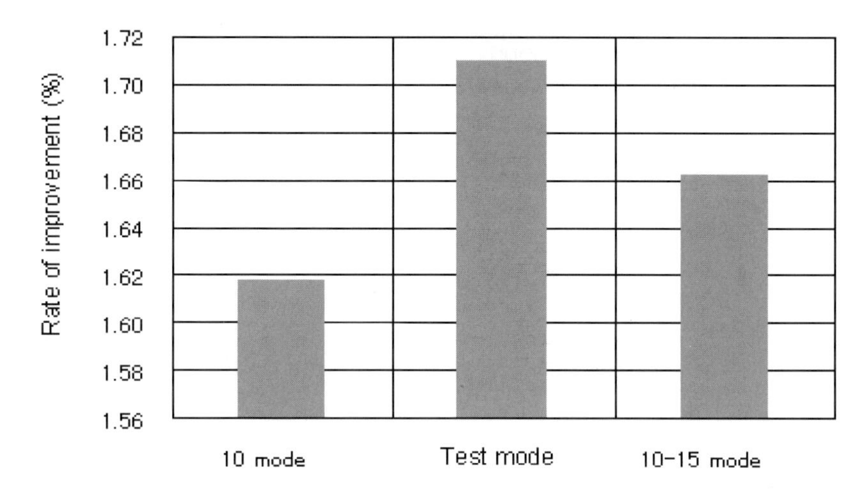

Fig. 8: Rate of improvement in test mode fuel economy

Fig.9 presents the difference of drive torque design between conventional system and integrated engine-CVT control system. As the conventional system is concerned, it is very difficult to obtain the linearity of actual drive torque through the limitation area like maximum and minimum mechanical ratio, maximum and minimum engine revolutions. And also change between torque converter region and lock-up region is an obstruction of linearity. When integrated engine/CVT control is used, target torque map is made smooth, and engine torque map is correct, drive torque can be made linear even in non-linear area of transmission.

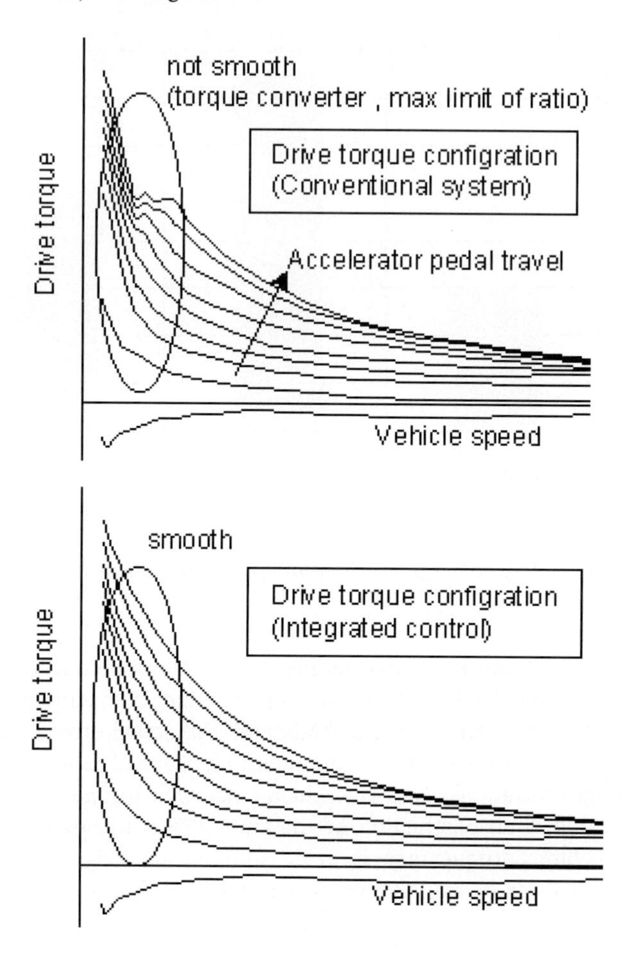

Fig. 9: Comparison of drive torque design

Improvement of Transient Drive Torque Response

Response of CVT is slower than conventional automatic transmissions because of continuous gear ratio control. In general, when a CVT-equipped vehicle is accelerated, it is said that there is no feeling of acceleration until the downshift ratio change has been completed. It has very hard technical difficulty and costs high to improve the response speed of CVTs, because of its mechanical limitation. However, it is easier to improve the response of drive torque, because drive torque is a product of engine torque and CVT gear ratio. Response of engine torque is much higher than CVT's. So the faster response of engine torque can compensate for the slower response of CVT. In the algorithm developed this time, target drive torque is divided by real CVT gear ratio to obtain the target engine torque as is said before. So that the target drive torque is realized automatically by increasing the engine torque when the CVT gear ratio changes slowly. Fig.10 plots the response of actual drive torque when the driver increase the accelerator pedal travel. When he increase accelerator travel from about 1/16 to 1/4 in a step way, it is clear that the change speed of real gear ratio is slower than its target. The target engine torque is higher than engine torque correspond to the accelerator pedal travel. As a result, real drive torque is generated in response near its target. Acceleration for drivers are comfortable at the same level as the response of CVT is improved. However, this algorithm is not for spreading the limitation of driving torque, the acceleration from higher load driving is less improved in response.

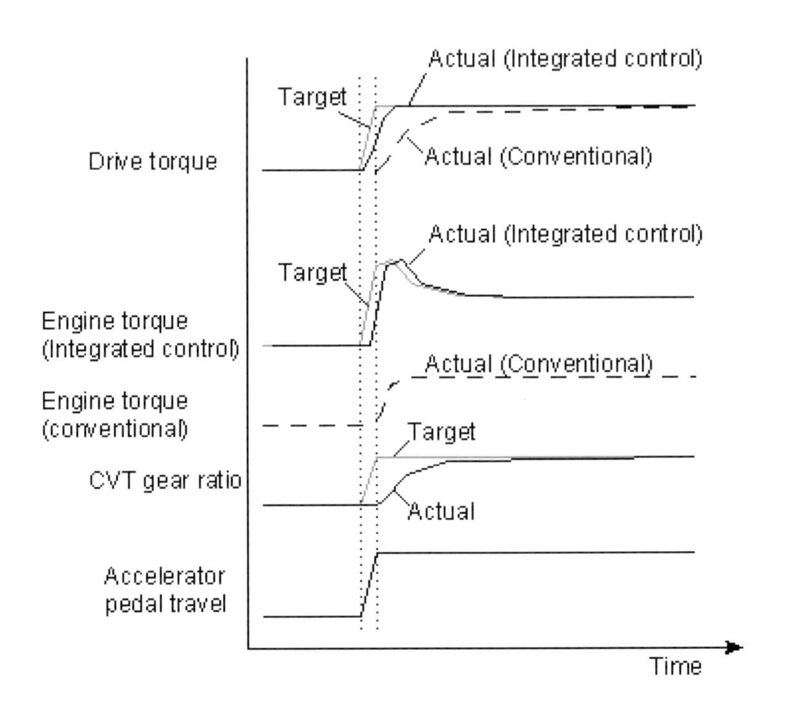

Fig. 10: Improvement of drive torque response

Arrangement of Communications for Advanced Vehicle Drive Control in the Future

Nowadays, traction control system and cruise control system are very popular as driving control systems which control vehicles through engine torque or driving torque. Vehicle dynamic control system has already sold as a product by some manufacturers. These systems usually control the lowest actuator units directly, so that control systems will become more complicated and generate more interference between algorithms when still more driving control systems are added in the future. Therefore it will be easier to avoid these problems by communicating through driving torque with driving control systems and controlling actuators with only final targets, even if the number of driving torque control systems will increase in the future. (Fig.11)

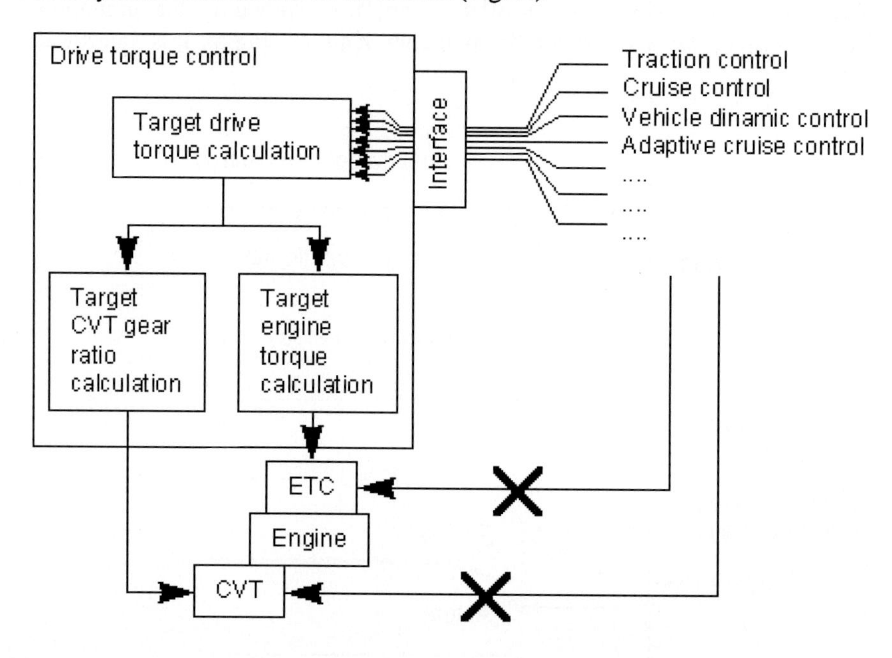

Fig. 11: Arrangement of Communications

Conclusion

1. A control algorithm has been developed for integrated engine/CVT control, making it possible to obtain the demanded drive torque with optimum fuel economy.

2. This algorithm enables powertrain system to have good response in drive torque, compensating CVT's slow response with engine torque.

3. This algorithm is a rational algorithm and has possibilities, considering drive control systems will become more multiplex and more complex in a few years.

Production of light wave conncectors, hermetic housings, heat sinks etc. by the MIM process

R.E. Hardt, Senior Partner, Industriekontor Rolf Hardt, Benrather Schlossufer 51, D-40593 Düsseldorf
Phone: ++49 211 712 828
Fax: ++49 211 713 186.

Keywords: Injection Moulding, MIM/CIM, Mass Production

By metal powder injection moulding (MIM) and ceramic powder injection moulding (CIM) near net shape components are produced of complex geometry with good surface finish and close tolerances. Through the use of much finer powders than commonly used in conventional sintering, parts are obtained with higher densities and enhanced mechanical properties with very uniform microstructure. The MIM technology aims at the production of complex, small components, difficult to produce satisfactorily and economically by known methods and from materials difficult to machine or otherwise to transform. Ideal component candidates are not only in new developments but also in existing products, by the combination of two or more parts into one single one and by avoiding expensive post machining and handling operations. Apart from fibre-optic components, hermetic housings, heat sinks etc, typical components are complicated small gears, parts for control and locking mechanisms, automotive fuel injection parts, automotive valve drain parts, airbag sensor parts, magnetic parts, computer disk drive parts etc. Materials range from NiFe, stainless steels, NiCo alloys, Ti alloys, WCu and MoCu to Alumina, translucent Alumina, Zirconia, Silicon Nitrides etc. Typical densities in the sintered form are from 93 to 100%. Surface finish 0.8 to 1.5 μ. Tolerances ± 0.3 to ± 0.2 % of nominal dimensions. Max weight approx 50 gr. CPK values > 1.45 on critical dimensions. MIM is a mass production process and offers best cost efficiency at high production quantities. It can also be an economical route for small quantities if parts are complicated in shape and cannot be made easily or economically by other manufacturing processes.

Advanced Packaging and Interconnection Technologies for Automotive Microelectronic Modules

Peter Sommerfeld, Daniel J. Jendritza, and Stephan Hell
Philips GmbH Mikroelektronik Module Werk Krefeld
Kreuzweg 60, D-47809 Krefeld, Tel./Fax: 02151 / 576-252 / -418

Keywords: Hybrid Circuits, Packaging, High Temperature Soldering

1 Introduction

The complexity of automotive electronics found in the vehicle electrical system with all its electro-mechanical functions has increased immensely in recent years.

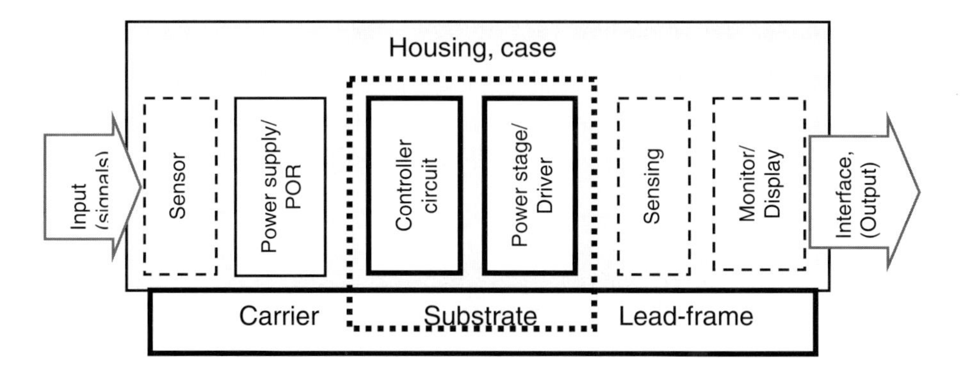

Fig. 1: Block Diagram for a typical automotive microelectronic module

In order to fulfil quality requirements in environments which are often very harsh, the trend has moved towards networks of electronic modules. Customer demand for increased levels of reliability, functionality, performance, and miniaturisation along with concurrent cost reductions requires an integration potential which can be achieved by a gradual combination of mechanical and electronic functions into actuators, sensors, switches, and controllers. This leads to sub-systems which are controlled by electronic central units for e.g. engine, transmission, and brakes [1].

This development has impacted on the very nature of automotive microelectronic modules. The packaging and interconnection technology used to build up these modules has been under severe scrutiny. It is no longer sufficient to focus on technologies and processes that centre around substrate fabrication and population, but instead it becomes necessary to cover peripheral components of these types of

modules as well. This can be achieved by having a broad technology base such that interconnection and packaging can be optimised for a specific product application.
Most modern automotive microelectronic modules can be characterised by a combination of several functional building blocks, as is sketched in Figure 1.
Many modules so far still only consist of controller circuit and power stage on a substrate as indicated by the bold dotted line, although the trend is clearly towards a full integration of all building blocks. It is the foremost task of interconnection and packaging technology to tie those blocks together, taking account of all requirements detailed above [1]. This paper will highlight some of the most important technologies used for the realisation of modern automotive microelectronic modules and give example for current products in high-volume production.

2. Substrate Technologies

2.1 Classical Thick-film Technology

Thick-film or hybrid circuit consist of screen printed layers of conductive, dielectric, or resistive materials on an alumina (Al_2O_3) substrate [2]. These layers are applied in form of an ink or paste, are subsequently dried and sintered at high

Fig. 2: A populated classical thick-film substrate

temperatures, where they form a strong bond with the alumina substrate. Thus a double-sided substrate with multilayers and integrated resistors is obtained, to which housed components can be soldered and naked dice can be bonded. These dice are covered by epoxy adhesive, or a so-called glob top, to avoid damage to the dice or bond wires. The connections to the hybrid circuit are made by single- or dual-in-line leads (SIL or DIL, see Figure 2). The whole module is coated with lacquer for protection. Hybrid circuits have a high level of integration and an excellent performance in harsh environments. That makes them ideal for applications in the automotive industry.

For conductor tracks gold, silver, silver-palladium, and silver-platinum pastes are used depending on requirements of soldering, wire bondability, and resistance. The thickness of conductive layers are in the region of 10 μm for a fired track. For extremely high conductance tracks a thick print is used. These can be realised up to a thickness of 80 μm giving a sheet resistance of 0.3 mΩ/□.
Dielectric layers are used to facilitate the formation of multilayer structures, as a barrier for solder, and as protection for the conductor tracks. They are used to form vias (connections) between active layers. This can be achieved by printing holes, which in a subsequent printing process, get filled with a conductor paste.
Resistors from 0.5 Ω to 10 MΩ can be printed with special pastes consisting of glass and precious metals. These resistors can be laser trimmed to a specific value or to adjust some functional parameter of a circuit. Temperature and long term stability are better than 1%. The temperature coefficient of resistance (TCR) of the resistor pastes is in the region of 100 ppm/K.

2.2 Direct Bonded Copper Technology

Direct Bonded Copper or DBC technology uses a ceramic substrate for bonding of thick copper foil of typically 0.3 mm thickness. The bonding mechanism is based on a thin layer of copper oxide on the foil which melts in eutectic composition with copper at a temperature of 1065 °C, slightly below the melting point of pure copper. A pre-oxidised copper foil on a Al_2O_3 ceramic substrate heated to that temperature will therefore wet the ceramic surface without losing its form stability. On cooling the copper oxide solidifies and ensures an intimate bond between foil and ceramic substrate [3].
Once the copper has been bonded to both sides of the ceramic conductor tracks can be defined by a masking and etching process. After cleaning the copper surfaces can be selectively plated by nickel and gold. Figure 3 shows a DBC substrate with gold-plated pads for bare die and component soldering.

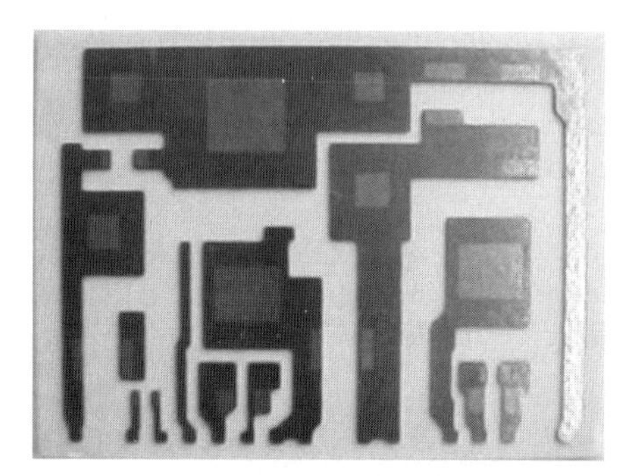

Fig. 3: A Direct Bonded Copper substrate with nickel- and selective gold-plating

A major advantage of DBC is the high current carrying capability. With the standard copper thickness of 0.3 mm sheet resistances of 0.06 mΩ/□ can be achieved. In addition, the copper causes a limited heat spreading effect, which in conjunction with the good thermal conductivity of Al_2O_3 reduces the thermal resistance to the heat sink. For even better thermal performance copper can also bedirectly bonded to aluminium nitride (AlN). One limitation of DBC is the conductor track pitch, i.e. the mid-to-mid distance between neighbouring conductors, which for standard copper thicknesses is around 1 mm.

2.3 Z-StrateTM Technology

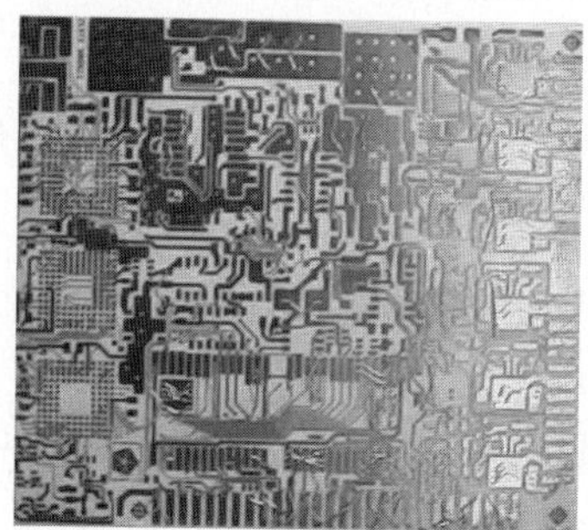

Fig. 4: Substrate in Z-Strate^{TM} technology

An new emerging substrate technology is called Z-Strate^{TM} (see Figure 4). Here, copper conductor tracks are deposited onto alumina substrates by a combination of electro-less plating, photo-lithography, electro-plating, and a strip/etch process [4]. The adhesion mechanism to Al_2O_3 is realised by an accurately controlled reduced oxygen firing process. Through-hole contacts to the flip side of the sample are also made from pure copper. Z-Strate allows very fine copper tracks (down to 50 μm) with the high aspect ratio of 1, i.e. the copper tracks can be made just as high as they are wide! The maximum thickness of the tracks is 125 μm. This gives Z-Strate ideal properties for high current applications with a sheet resistance of the conductor tracks of 0.14 mΩ/□. Z-Strate has a slightly better thermal properties than hybrid thick film because of the thick copper, that can have heat-spreading ability. Z-Strate is an excellent choice of substrate when high current and fine pitch are required.

This new technology, developed by PG Zecal (USA), is being qualified at Philips Microelectronic Modules at the moment in close collaboration with the Philips Centre for Manufacturing Technology.

3. Interconnection Technologies

In order to integrate power devices onto one substrate, very often unpackaged bare dice are used. This allows a miniaturisation of the function and an optimised thermal management, as will be discussed further below. An important technology

Fig. 5: Bare, wire-bonded power dice on a hybrid thick-film module

for the interconnection of such dice is thick wire bonding, where aluminium wires with diameters from 100 μm to 500 μm are attached to the die and the bond pad on the substrate by means of ultrasonic bonding. Figure 5 shows an H-bridge driver module for seat adjustment motors with wire bonded power transistors. The current which has to be carried determines the diameter and the number of wires for the interconnect.

Figure 6: Control IC as bare die with Au thin wire bonds

Bare dice which are controller ICs do not have high current requirements, but usually a high number of I/Os and small pads for wire bonding. The interconnects between substrate and die are therefore achieved by thin wire bonding with wire diameters between 25 μm and 50 μm, see Figure 6. Two slightly different technologies are used: thermosonic bonding of thin Au wires and ultrasonic

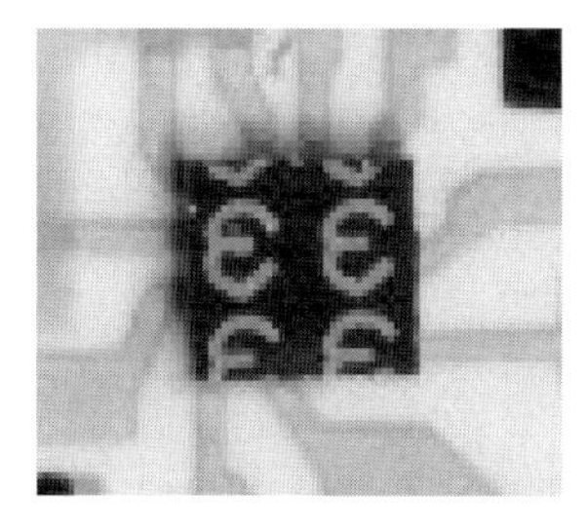

Fig. 7: Flip chip on a ceramic substrate

bonding of thin Al wires. For all types of wire bonding it is important to have bond pads with the right kind of metallisation. The mechanical properties of the

substrate material can also strongly influence bond behaviour and yield [5]Alternatively to wire bonding the controller dice, some semiconductor suppliersalso offer bare dice with flip chip connect, see Figure 7. Here, small solder bumps are deposited onto the connector pads on the active side of the die. The die can be attached to the substrate by placing it "flipside up", i.e. with active layer down. After a reflow soldering process the bumps will form the interconnect to the substrate [6]. Since the flip chip is used for an under-the-hood product, high temperature solder is used for both bumping and the solder paste for reflow.
The interconnection from the substrate to the next package level, which is usually a plastic housing, is achieved by either thick wire or ribbon bonding, or solderedleads. The most reliable types of soldering technology are solder reflow

Fig. 8: Bare power dice soldered onto heat-spreaders

and micro-flame soldering. The latter allows the leads to be integrated into the housing but is a sequential process [7]. The former is a parallel process, but the leads cannot be part of the housing because the plastic typically will not withstand the reflow temperatures, since the high temperature solder is used for automotive applications with maximum ambient temperatures of 140 °C.

4. Packaging Technologies

Several types of packaging steps have to be applied to obtain a reliable and compact module. Power dice are usually soldered onto a small metal plate (copper or compound) larger than the die (see Figure 8). This plate serves several purposes and acts as:

- a lateral thermal short to prevent the occurrence of hot spots mainly in bipolar devices
- a heat capacitance to absorb short power spikes and therefore limit the junction temperature
- a heat-spreader, which distributes the heat generated in the device over a larger area and thereby reducing the junction temperature

Fig. 9: Bare power dice with glob top encapsulation protecting both die and wire bonds

Fig. 10: A module with a transfer moulded encapsulation

The soldering process has to be fluxless so as to ensure a low void level and therefore a low thermal resisitance to the substrate. In order to withstand the high temperatures found in automotive environments high temperature solder is used. Bare dice have to be protected in some way to ensure that the interconnections are not damaged in some way. This can be achieved by a so-called globular top (or glob top), where adhesive is dispensed onto the die and cured. The adhesive has a high filler content in order to match the thermal expansion coefficients and not damage the die or interconnections. If a plastic housing is used, potting with silicone gel is an effective way of protecting the whole module. The soft gel imposes no mechanical stress onto the interconnections during temperature changes and protects the circuit from environmental influences. A plastic cover can also be fitted to protect from mechanical damage by larger particles or seal the circuit cavity completely. Another technology which offers excellent protection of the circuit and components is transfer moulding, which is widely used for the packaging of ICs. High investment, tooling costs and restrictions for the moulded geometry make this technology very application specific. A plastic housing is probably the most versatile way of realising the final package level of an automotive microelectronic module. (See figure 11)

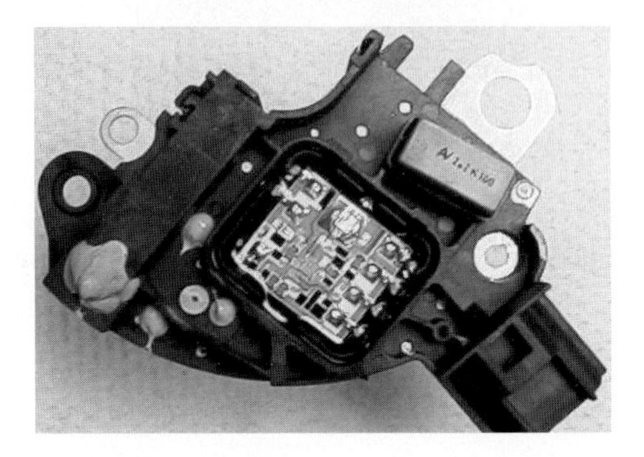

Fig. 11: Automotive microelectronic module in a plastic housing

It can integrate a leadframe, odd components like larger capacitors and inductors, and provide electrical and mechanical interfaces like plugs, brushes, and points of fixation [8]. This leads to a level of integration which often referred to as mechatronics.

5. Conclusions

There is a strong demand for highly compact and integrated automotive microelectronic modules. The interconnection and packaging technology for these modules relies greatly on advanced substrate, wire bond, soldering, encapsulation and housing technologies. Philips Microelectronic Modules has addressed these issues by offering a technology portfolio that goes far beyond classical thick film technology. These technologies, e.g. high temperature processes, have reached a sufficient maturity to fulfil the demanding reliability requirements for automotive products. In future it will be even more important to be able to offer a wide range of interconnection and packaging technologies to meet the demands of ever increasing miniaturisation, integration, mechatronic diversification, reliability, and cost pressure.

References

[1] Appel K. and Jendritza D.J., *Demands of the car industry on future microelectronic products,* VTE 10 (1998) Heft 2, DVS-Verlag GmbH, Düsseldorf, pp. 78 – 83.

[2] *Thickfilm and Diffusion Patterning process features*, Fact Sheet 02, Philips GmbH Mikroelektronik Module, Krefeld.

[3] Waibel, B., Martin, W., *DCB-Substrate für die Leistungselektronik - Eigenschaften und Anwendungen,* Zuliefermarkt 3, 1990.

[4] *Advanced Z-Strate™ Substrate Technology*, Fact Sheet 05, Philips GmbH Mikroelektronik Module, Krefeld.

[5] Thor V., Engbring J., Nover M., and Sommerfeld P., *Manufacturing a BGA Multi-Chip-Module,* Proceedings of EuPac '98, Nuremberg.

[6] Thor, V., Appel, K., Nover, M., and Sommerfeld, P., *Soldered Flip Chip ICs in automotive applications,* Proceedings of the IEMT '98, April 27 - 29, Berlin.

[7] Engbring J. and Jendritza D.J., *Gehäusemontage von Keramikhybriden mit Hilfe eines Flammlötprozesses,* VTE 10 (1998), DVS-Verlag GmbH, Düsseldorf, pp. 130 – 135.

[8] *Alternator regulators - Technology for the harsh automotive environment*, Fact Sheet 03, Philips GmbH Mikroelektronik Module, Krefeld.

A low cost, fully signal conditioned pressure sensor microsystem with excellent media compatibility

Roy Grelland, SensoNor asa, P.O. Box 196, 3192 Horten, Norway
Henrik Jakobsen, SensoNor asa, P.O. Box 196, 3192 Horten, Norway
Bent Liverød, SensoNor asa, P.O. Box 196, 3192 Horten, Norway

Abstract: In order to make cost efficient systems for pressure measurement there is a need for pressure sensors that meet certain requirements:

- Low cost, high volume manufacturing.
- Mountable on PCB's by automatic pick and place machines on high volume assembly lines.
- The sensor element should be robust and not be influenced by dust and other contamination in the pressure media to avoid additional membranes or other expensive, complicated protections. No need for external components for signal conditioning and calibration.

SP16 [1] is a transfer moulded pressure sensing microsystem with internal signal conditioning ASIC and a unique triple stack sensor element that meets these requirements.

Keywords: Pressure Sensors, Packaging, Media Compatibility

1. Environmentalcompatibility through triple stack-sensor element

The sensor element, made from single crystal silicon, is sandwiched between two glass layers (fig. 1). When exposed to external pressure, four piezoresistors forming a Wheatstone bridge will sense a mechanical stress in the silicon diaphragm. The resistance value will change according to the piezoresistive effect in doped silicon.

Fig. 1: Triple stack sensor element

The sealed vacuum cavity is made by anodic bonding. Anodic bonding consists in joining the silicon wafer with a borosilicate glass wafer, such as Pyrex #7740. The anodic bonding is achieved by putting the silicon and the glass wafers on a hot plate (300 - 450 °C) and applying a high *DC* voltage to the stack (800 - 1000 V), such that the glass wafers are negative with respect to the silicon. At the bonding temperature, the alkali oxides present in the glass dissociate in positive ions, primarily sodium Na^+, and negative oxygen ions. Under the action of the electric field, the very mobile sodium ions leave quickly the region near the anode and a depleted layer about one micron deep is formed. The high electrostatic forces which develop across this layer move the non-bridging oxygen ions and direct them towards the silicon anode, where they oxidise the silicon and form strong silicon-oxygen bonds. The wafers are polished to a smooth surface and carefully cleaned prior to the bonding step.

By using diffused and buried conductors [2], the piezoresistors and the interconnection leads are placed on the same side of the sensor element as the vacuum cavity. The pressure inlet is through an opening in the bottom glass layer, which is also attached to the silicon layer in the same anodic bonding process as is used to make the vacuum cavity. The great advantage of this technique is the resulting improved media compatibility when having the pressure inlet port towards the backside of the silicon diaphragm, and not against the much more corrosion and moisture sensitive front side with metal conductors, piezoresistors and the electrical interconnections. The result is a sensor element with excellent media compatibility.

Figure 2 shows the special patented feed-through method that is used to form the sealed vacuum cavity and at the same time to form the electrical connections passing the anodic bonding area.

Fig. 2: Feed-through for vacuum cavity

Within an n-well, several p-type-diffused islands are realised for each required crossing. An n-epi layer is grown on top in order to bury the p-conductors. Highly boron doped layers are implanted and diffused through the epitaxial layer in order to contact the buried conductors. A shallow n^+ layer is also implanted right under the anodically bonded areas in order to minimise the electrical effects of the silicon-glass interface on the junctions underneath.
The quality of the bond can be evaluated on three main criteria: void density, bond strength and thermal residual stress. The mechanical strength of the anodic bonded vacuum reference volume as measured by die-share technique is in the range 20-50 MPa. The quality of the enclosed vacuum inside the cavities has been monitored for different bonding conditions in order to optimise the process towards achieving the minimal residual gas pressure. A great advantage of using glass is the possibility for seeing through the glass making it possible to inspect the hermetic seal between the glass and the silicon. Unbonded areas like voids and leaking channels caused by particles or scratches on the surfaces before bonding can be seen through the glass in a microscope, hereby giving the possibility to inspect for and reject no-good dice with potential gas leakages.
The diaphragm thickness is defined by etch-stop against pn-junction using the 4-electrode configuration [3]. Very good thickness control is obtained by using this method. The method is suitable to form diaphragms with thickness in the range from about 3 micron up to more than 25 micron. Typical thickness control obtained by using this method is in the range +/- 5%.

2. Low cost, high volume manufacturing suitable for automatic pick and place through specially developed transfer molded package

Transfer molding [4] is recognised as being a highly efficient method for packaging of high volume components in the integrated circuit industry. SensoNor has established an assembly line for microsystems based on the standard processes for assembly of integrated circuits (see fig. 3). The line consists of standard processes such as die attach, wire bonding, transfer molding, tin/lead plating and trim/form.

Fig. 3: SensoNor´s assembly line for microsystems

SP16 is made with a standard copper lead frame. SP16 consists of two dice; a sensor die and an ASIC. The ASIC is a standard CMOS die, and no special die attach is required. The sensor die is as explained in section 1, a three layer structure with an opening in the bottom glass layer to let the pressure media in to the membrane. In order not to fill the interior of the sensor die with die attach epoxy, a special programmable dispenser is used to apply the die attach epoxy on the lead frame. By dispensing a carefully designed pattern, die share values in the range of 11 kg is achieved without any epoxy entering the interior of the sensor die. (Production data shows an average die shear value of 10.84 kg, and a Cpk of 4.32)

Fig. 4 : SP16 Die to die wire bonding

Since SP16 consists of two dice, die to die wire bonding is used. Ball-wedge wire bonding is used with gold wires and aluminium bond pads (see fig.4). As the sensor die is a three-layer device, whilst the ASIC has one layer only, the wire bonding between sensor element and ASIC is done over a height difference of ~500 μm. This causes some problems w.r.t focusing of the camera in the automatic wire bonder, but by careful design of the wire bonding system, a stable wire bond process with a capacity of > 2000 units/hrs has been achieved.

The most demanding process in the automatic assembly line for SP16 is the transfer molding [4]. Although transfer molding is a standard process in the integrated circuit industry, transfer molding of sensor components gives some special challenges. The transfer molding is performed with liquid epoxy at ~180°C and ~100 Bar. The challenge is of course to avoid that the interior of the sensor is filled with epoxy during the transfer molding process. During the process development, a lot of different mold designs were tried. In Taguchi experiments the different mold designs were tried together with different process parameters.

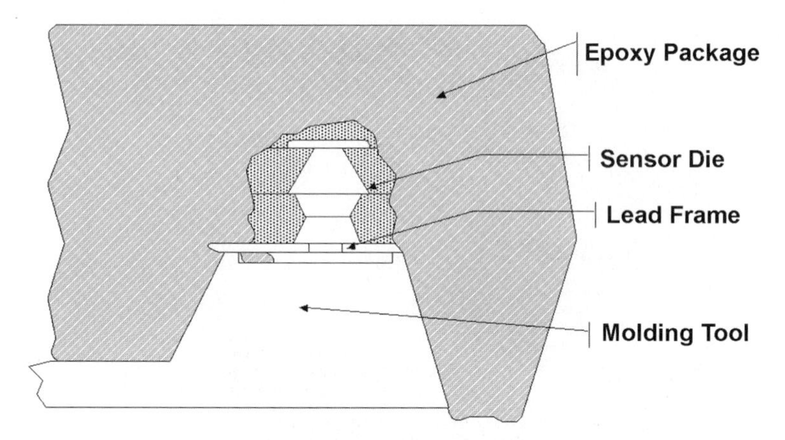

Fig. 5: Molding tool detail.

As can be seen from fig. 5, the idea of the mold design is to block the area around the hole in the lead frame against the epoxy. The solution was found in that the outer circumference of the area to be blocked, were clamped with a tight clamp. In the area on the inside of the clamp, some room is left between the molding tool and the lead frame. This solution works in the way that the liquid epoxy will indeed penetrate under the clamp on the outer circumference of the blocked area. But because there is more room just inside the clamp, there is no more capillary effect to help the continuous flow of the molding compound towards the hole in the lead frame. The result is that the flow stops just inside the clamp, and the molding compound cures there

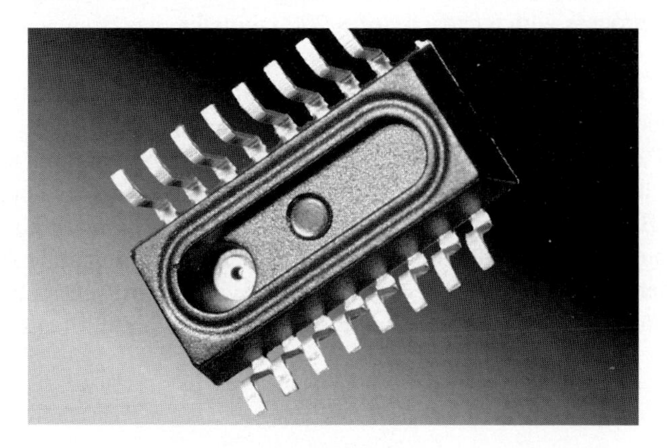

Fig. 6: Package with opening for pressure inlet at the bottom side.

For efficient mounting on printed circuit boards, an additional requirement for the SP16 package is that it should be mountable by standard pick and place robots. By making the opening for the pressure inlet on the bottom side of the component, the top surface is left flat and ready for surface mounting. Placed on a PCB with a hole, the component is ideal for measurement of pressure where the component is placed inside the media that it will measure.

For remote pressure measurements, a special mechanical arrangement is required.

Fig. 7: Transfer molded component easily mountable by standard pick and place robots.

3. Component performance with internal signal conditioning

Due to the spread in the silicon micromachining processes, each sensor die has a somewhat different characteristic with respect to sensitivity, offset and temperature drift of the same. Therefore, each sensor device has to be individually

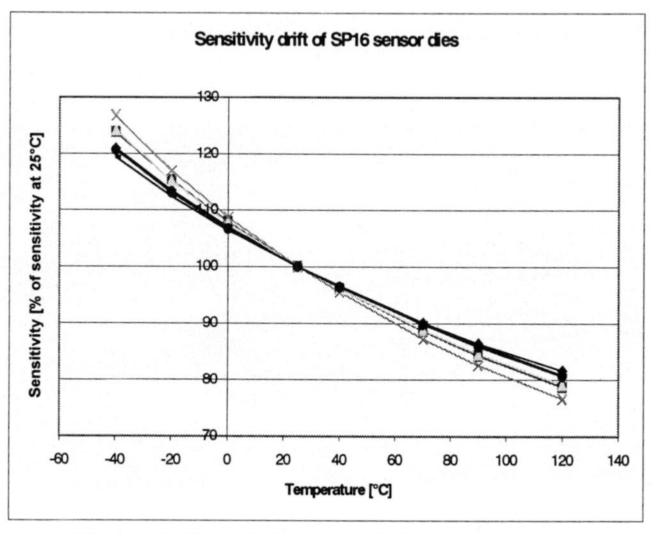

Figure 8: Sensitivity drift of SP16 sensor dies

calibrated during production. It is assumed that the temperature drift of the sensitivity and offset can be modelled using 2. order models, and the pressure sensitivity is assumed to be linear. The following figure shows the sensitivity drift of a typical set of SP16 sensor dies.
The total sensor model can then be expressed as follows:

$$v(p,T) = \frac{s0 \cdot p + z0 + z1 \cdot (T-25) + z2 \cdot (T-25)^2}{1 + s1 \cdot (T-25) + s2 \cdot (T-25)^2}$$

where: v(p,T) is the output signal in mV/V
s0 is the sensitivity at 25°C
s1 and s2 are 1st and 2nd order coeffisients of the sensitivity drift
z0 is the offset at 25°C
z1 and z2 are 1st and 2nd order coeffisients of the offset drift
p is the pressure (in bar)
T is the temperature (in °C)

This model is hard-wired into the digital part of the ASIC, and thus only the parameters can be changed, not the model itself. During production testing, each of the parameters describing the specific sensor die, is programmed into an on-chip OTPROM. This is being done in an automated, purpose designed test line. This means that no additional trimming, filtering or software offset is required, nor are any external components. The following figure shows a block diagram of the ASIC, including the OTPROM and the compensation unit, containing the sensor model

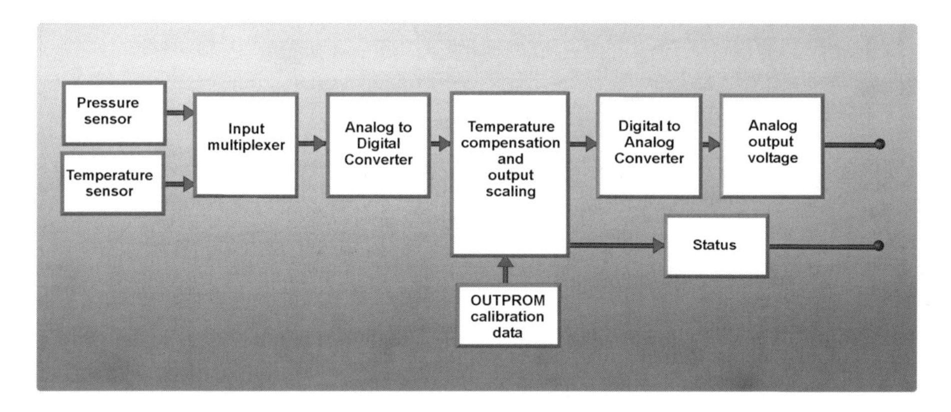

Fig. 9: Functional block diagram.

One of the challenges is to test the devices in such environments (pressure/temperature) that an accurate set of parameters can be obtained, yet maintaining the yield and capacity normally expected by a high volume/low cost product.

The following plot shows typical measurement performance for the SP16 at 1000hPa over the temperature range –40°C to +125°C.

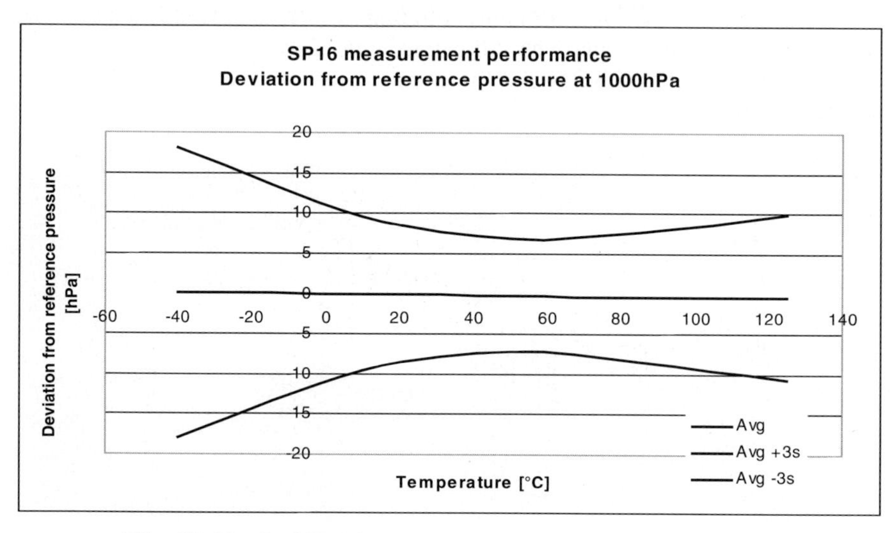

Fig. 10: Typical SP16 measurement performance

4. Component testing and qualifiacation

The SP16 has been subjected to a qualification test program that includes several environmental exposures. Of these can pressure cook (autoclave), high and low temperature storage, gas test, liquid immersion test, thermal shock, life test and vibration tests be mentioned. The SP16 has proven to withstand all these harsh environments successfully. As an excample, the following plot shows the average change in measurement accuracy after being subjected to the pressure cook test. Pressure cook test is considered to be a particularily harch test for this kind of components.

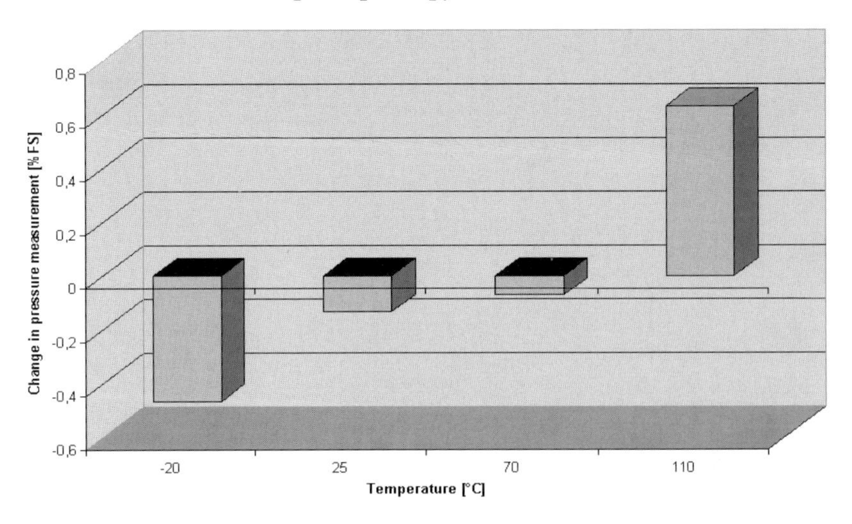

Fig. 11: Pressure Cook Test results

This shows that the nature of the design of the SP16 pressure sensor makes it very robust with respect to contamination from dust and other pollution in the pressure media.

5. References

[1] SensoNor asa, Data sheet: SP16 general purpose pressure sensor, SensoNor asa, Horten, Norway, 1998

[2] A, Cozma, H. Jakobsen, R. Puers, Electrical Characterization of the Anodically Bonded Wafers, J. Nucrinecg. Microeng, *to be published 1998.*

[3] D. Lapadatu, G. Kittilsland, M. Nese, S.M. Nilsen, H. Jakobsen, A Model for the Etch Stop Location on Reverse Biased pn Junctions, Sensors & Actuators A, 60-63, 1998.

[4] Louis T. Manzione, Plastic Packaging of Microelectronic Devices, p. 182-243, 1990.

Occupant Classification System for Smart Restraint Systems

K. Billen, L. Federspiel, B. Serban
I.E.E. International Electronics & Engineering, Luxembourg

Abstract: The primary function of the Occupant Classification System is to provide reliable passenger seat occupancy information to the automobile's central processing unit to control airbag deployment..

Our Occupant Classification (OC) sensor system is based on analysis of the seat occupancy pressure profile, which discerns human like from human unlike profiles. If the occupancy is identified as a person, an allocation into one of four morphologic ranges is made. Accordingly, children, light adults, heavy children, medium adults, heavy adults, etc. can be discerned

Keywords: Force Sensing Resistor, Occupant Classification, Smart-Airbag-System

1. Introduction

In contrast to actual one-stage airbag systems, future restraint systems will offer a multitude of triggering possibilities. The aim of such systems is to reduce the risk and level of injuries by automatically adapting the airbag deployment and seat belt pretensioner to the driving status of the vehicle, its occupants and the crash severity.
Several systems are in development. They are based on different measurement principles, such as strain gauges, liquid pressure, optical fibers, capacitive, ultrasonic, infrared, etc. IEE is developing a sensor based on FSR® (Force sensing resistor) technology. The sensor is assembled in the seat and captures a pressure related seating profile.

Literature and tests performed with hundreds of persons indicate a correlation between the anthropometrical characteristics of the person and its corpulence. The analysis of characteristic parameters extracted from the pressure profile allows classification of seat occupancy (Figure 1).

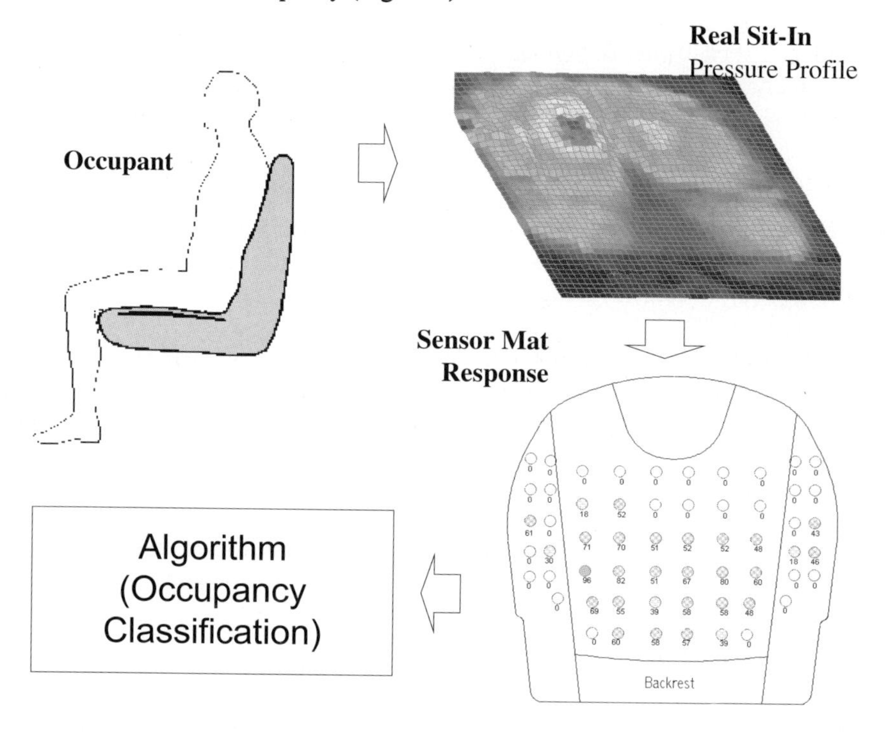

Figure 1: OC principle

2. Description of the proposed system

2.1 System Overview

The total system consists of the following sub-systems (Figure 2):

OC Sensor mat (FSR® technology)

The sensor mat consists of two polymer films screen printed with conductive patterns and connecting leads, then sandwiched together with a polymer spacer sheet in between (Figure 3). Up to 100 single cells can be integrated in one sensor mat. A special circuit design provides access to each cell.

Measurement unit (Electronics unit)
With an ASIC controlled by a microcontroller, all FSR®-cells are measured and their resistance values are digitized. A specific algorithm is used handling the digital resistance matrix to provide an object classification that is used by the airbag control unit.

Airbag control unit
This unit controls the deployment of the airbag using all available crash information.

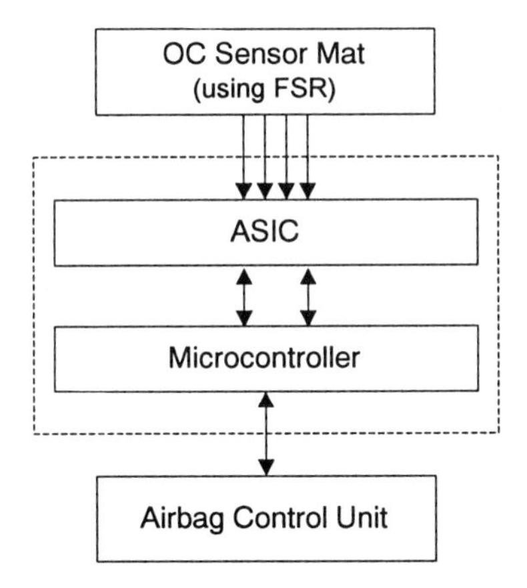

Figure 2: System Overview

2.2 FSR® technology

The Force Sensing Resistor® sensor consists of two facing substrates on which proprietary pressure-sensitive films and electrodes are printed (Figure 3).

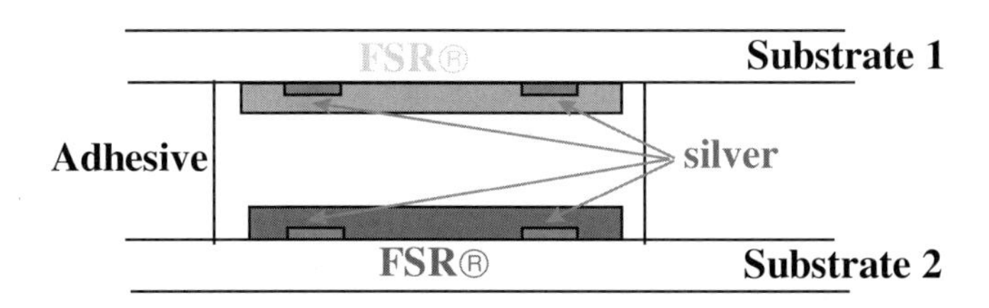

Figure 3: FSR® technology

When a force is applied on a sensor cell its electrical resistance decreases. This decrease is a monotonous function of the applied force.

2.3 Sensor mat

The occupant classification sensor consists of a mat with multiple sensor cells whose resistance values are acquired and converted into a set of digital values which provide a discrete pressure profile (Figure 5)..

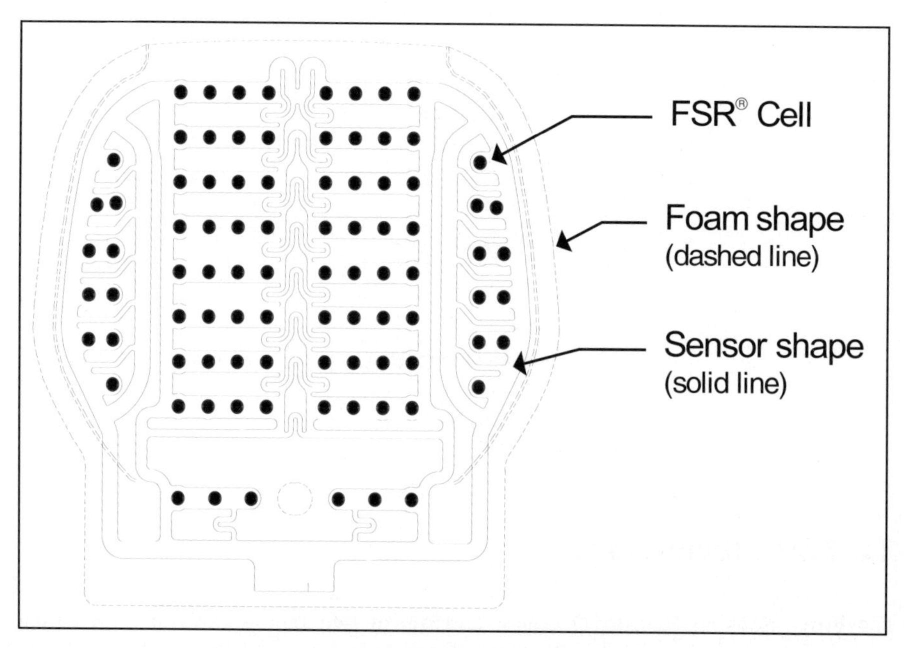

Figure 4: OC Sensor shape

The sensor is assembled between the seat foam and cushion. Each unique seat design requires specific sensor adaptation. Since 1995, starting with the Passenger Presence Detection System, IEE has adapted its sensors to more than 50 seat types. Figure 5 shows an equivalent circuit of the OC mat. The FSR® cells (R) are connected by printed silver leads (R_l) in a matrix configuration. Fixed resistors (R_f) at the end of each row and column connection allow for several hardware self-tests, e.g., short circuits, wire breaks, leakage current, etc. to check the sensor mat integrity.

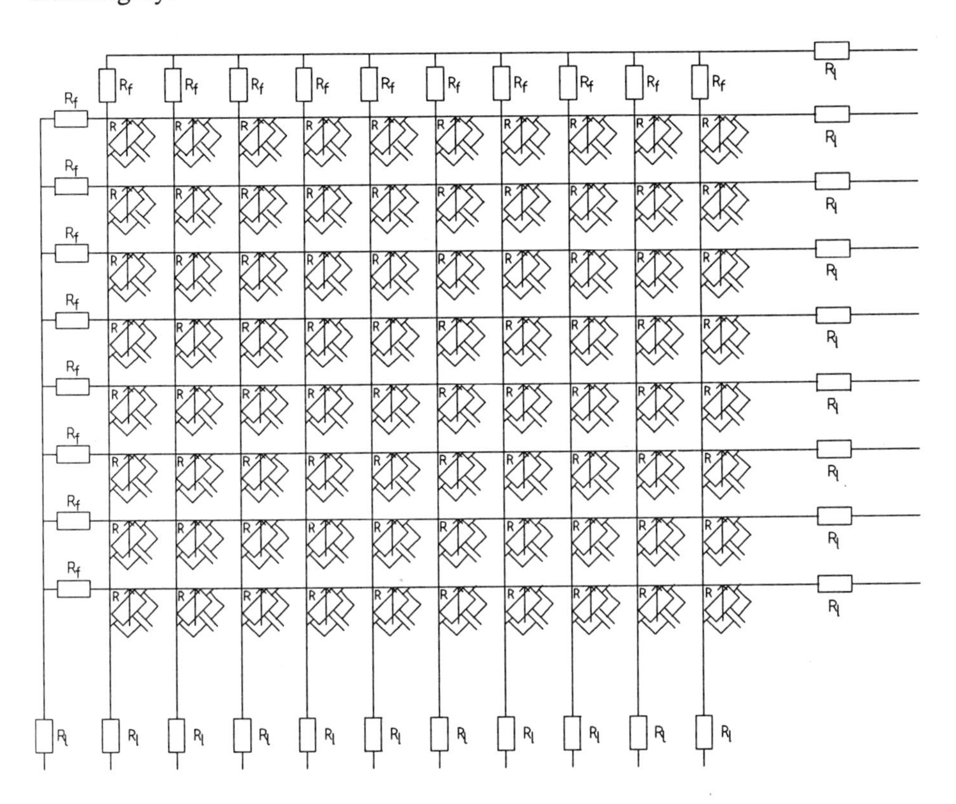

Figure 5: OC Circuit Diagram

2.4 Measurement Unit

This electronics unit consists of four main components: the microprocessor, the ASIC, the EEPROM and the communication interface.

The microprocessor controls all operating sequences and runs the classification algorithm.

The ASIC is the unit that interfaces the FSR® cells with the microprocessor. All the measurement steps performed by the ASIC are controlled by the microprocessor via serial synchronous interface.

The EEPROM stores all the configuration data and relevant system data in case of a crash.

The communication interface allows the right data flow between the OC electronics unit and the airbag control unit according to the OEM protocols. Therefore, every OC electronics unit is customer dedicated.

2.5 Data processing

The processing of the data from the acquisition to the classification is shown in Figure 6. The discrete pressure profile is analysed by an algorithm, which evaluates several parameters and calculates the digital pressure data. Different filters and memory functions make the system insensitive to the dynamic environment and movements of the seat occupancy. The mathematical and logical evaluation of all the calculated parameters determines the classification (airbag control data).

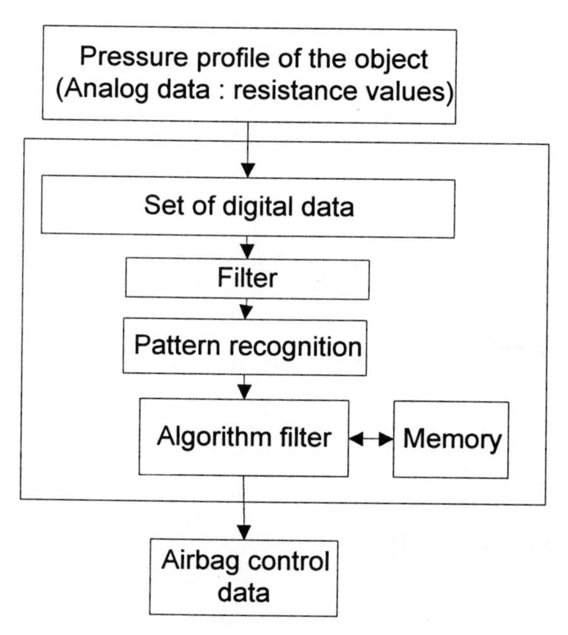

Figure 6: Data processing

The primary function of the OC-Algorithm is the classification of objects. First, we make a distinction between human beings and non-human beings, then further distinguish according to human anthropometric characteristics. Typical objects of the non-human class are child seats, cases or bags.

2.6 Main Profile Parameters

The complete occupant identification and classification is made through the evaluation of specific functions, called *Profile Parameters*. The most important ones are described below:

The ***Object Parameter*** is used to discern human like and human unlike occupancies. This parameter relies on the fact that the interaction between the human buttock and the seat surface leads always to a compact pressure profile while all other types of occupancies lead to non-homogenous profiles.

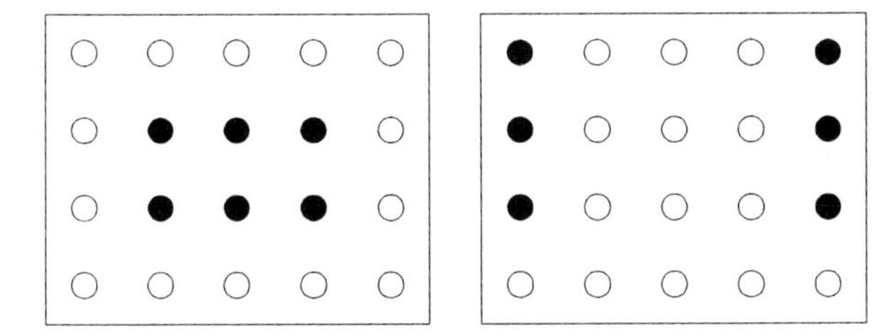

Figure 7: Object Parameter

Figure 7 depicts two distinct types of pressure profiles, both cases have only six activated sensor cells. A very compact activation pattern (left) is typical for human beings, while child seats usually generate a dissipated, fuzzy pattern (right). The *Object Parameter* is able to make a clear distinction between these two patterns by analysing their topological distribution despite having the same amount of activated cells in both patterns.
Figure 8 shows the calculation of this parameter for different people and child seats. It can be seen that at the same weight (e.g. at 40kg in the graph) the *Object Parameter* may have different values if the occupant is a child seat or a human being.

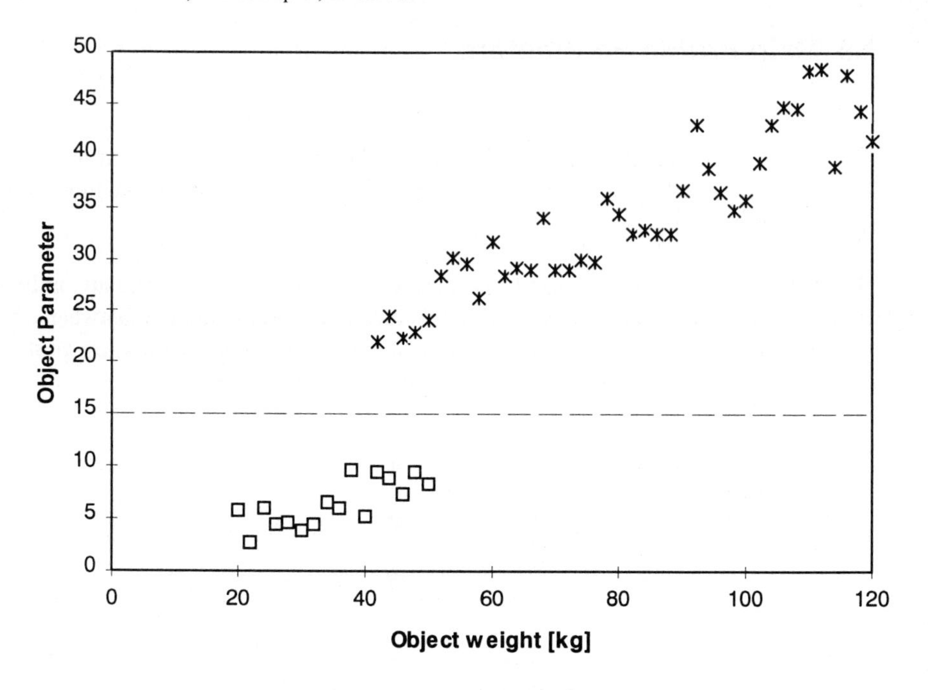

Figure 8 Object Parameter (* Humans; □ Child Seats)

The ***Coherence Parameter*** is an additional parameter used for the profile identification. It is used to evaluate the size and the shape of an activation pattern. An additional information concerning profile discontinuities is also provided.

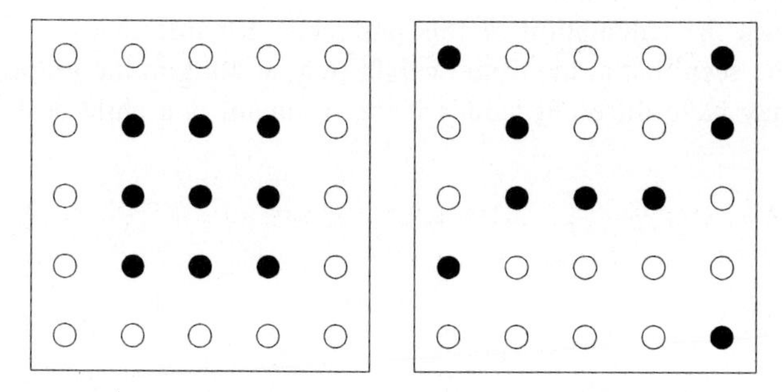

Figure 9 Coherence Parameter

Figure 9 shows the role of the *Coherence Parameter* in the occupant classification. Again, an equal number of cells are activated in both cases. In the left profile the cells are forming a coherent area, while they are distributed in a discontinuous manner in the right (non-human) profile. The profiles are easily distinguishable one from the other by the absolute value of this parameter.
The *Coherence Parameter* has a similar behaviour as previous but depends more significantly on the weight of a human being as it is shown in Figure 10.

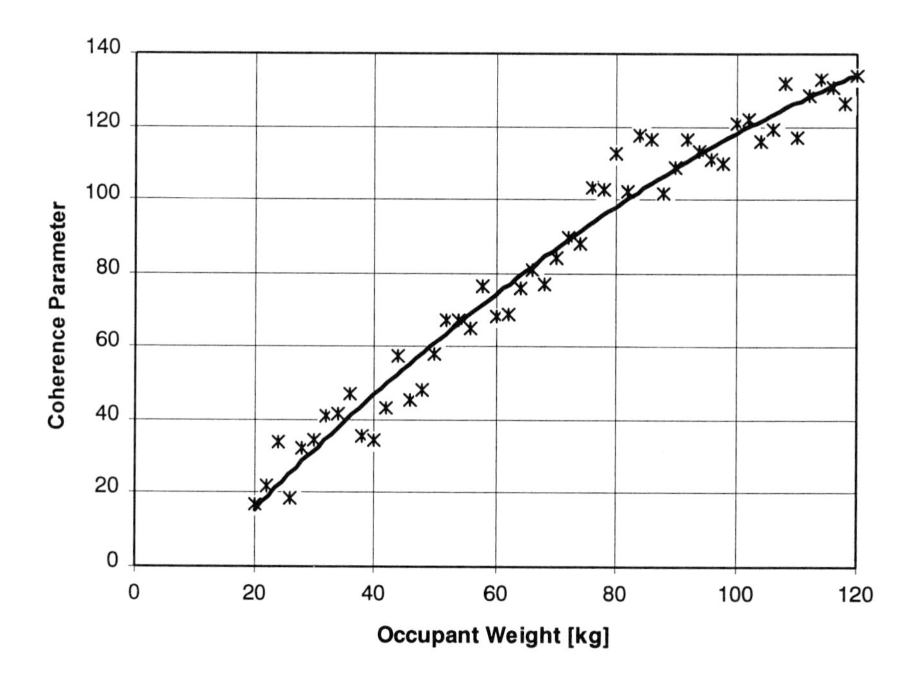

Figure 10 Coherence Parameter (Human Beings)

The third parameter, ***Width Parameter,*** allows classification of the human passenger by major anthropometric class.

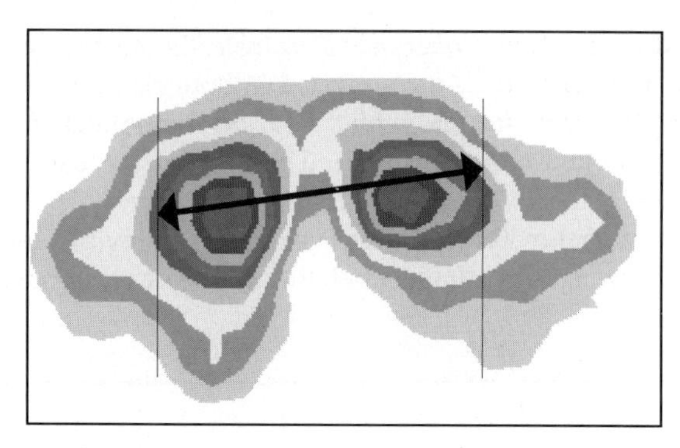

Figure 11: Width Parameter (Humans)

As shown in Figure 11, the Width Parameter contains the information concerning the geometry of the buttock pressure profile.
Figure 12 shows the dependency of the *Width Parameter* with weight from human beings. According to the actual implementation of the algorithm, the Width Parameter will not be calculated if the occupant is not a human being.

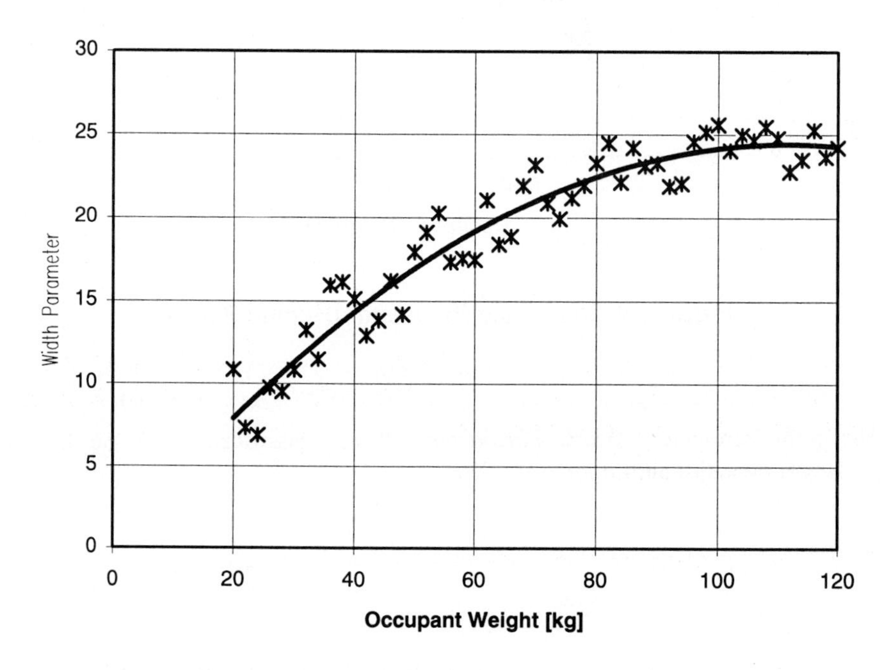

Figure 12: Width Parameter (human beings)

The ***Profile Quality*** function is the parameter, which allows a stable classification during the seat occupancy. It quantifies the overall profile quality or *goodness* by combining the previously defined parameters and seat specific parameters. Its evaluation provides critical information to the filtering mechanism. Thus, by using the information about the object position, the activation pattern, the profile orientation and by following a pre-programmed decision tree, the filtering procedure is able to provide at any moment the correct passenger classification with no regard to the instantaneous passenger position on the seat.

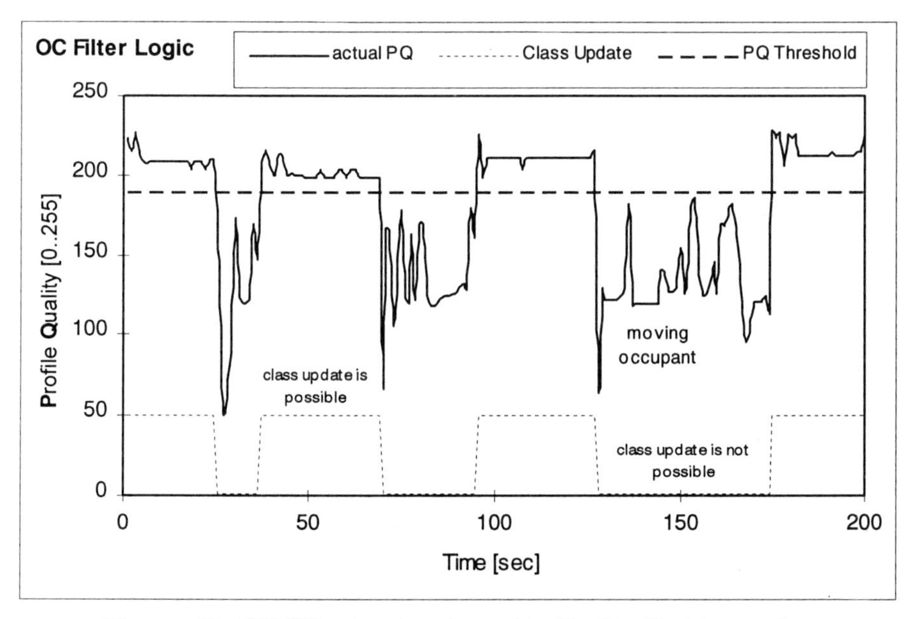

Figure 13: OC Filter Logic using a *Profile Quality Parameter*

Figure 13 shows how the classification update is made when the Profile Quality has values higher then a pre-determined threshold. This Threshold is itself a function of history and occupant position.

3. Conclusion

The morphology-based occupant classification system is able to provide a reliable, repeatable and instantaneous classification. The system discerns efficiently a child seat and other non-human seat occupants from the human ones. The human occupants (passengers) are classified with regard to their corpulence and to other anthropometrical proprieties. A precise position on the seat of their center of gravity is also provided.

Acknowledgment

The authors would like to express their gratitude to the OEM's and organizations who are supporting I.E.E. in the development of the Occupant Classification system.

References

The BMW Seat Occupancy Monitoring System: A Step Towards "Situation Appropriate Airbag Deployment*, Klaus Kompass, Michel Witte, SAE Paper

Automatic Passenger Presence Detection and Child Seat Orientation Detection, Andreas Hirl, Peter Popp, Joachim Uhde, Paul Schockmel, SAE Paper

Situation Appropriate Airbag Deployment: Child Seat Presence and Orientation Detection (CPOD), Thierry Goniva, I.E.E. S.àrl.

Real World Development of Adaptive Restraint Systems through the Use of Anthropometrically scaled Occupant Simulation Models,
L. Michaelson, R. Hoffmann, Alzenau

Anthropologischer ATLAS, Alters- und Geschlechtsvariabilität des Menschen,
Dr. B. Flügel, Dr. H. Greil/Prof.K. Sommer

1998 Anthropometric Survey of U.S. Army Personnel: Methods and Summary Statistics
United States Army Natick Research, Development and Engineering Center, Natick, Massechusetts 01760-5000,
C. Gordon, Th..Churchill, Ch. E. Clauser,
B. Bradtmiller, J.T. McConnville, I. Tebbetts, R. A. Walker

Occupational Ergonomics, Theory and Applications, James F. Annis and John T. McConville edited by Amit Bhattacharya, James D. McGlothlin

Body Dimensions of People; Terms and Definitions, measuring procedures, DIN 33402, Part 1

Concept for intelligent rear light aiming for compensation of environmental effects

W. Robel[1], G. Gruhler[2], D. Haas[2], P. Hage[3], P. Heelan[3], J. Apitz[4], J. Hewit[5], L. Pattison[5],

[1] Reitter & Schefenacker GmbH & CO.KG, Germany
[2] Steinbeis-Transferzentrum Automatisierung (STA) Reutlingen, Germany,
[3] JL-Automation, UK
[4] Jenoptik L.O.S., Germany,
[5] Dundee University, UK

Abstract

The joint European project (ESPRIT IntACT) presented outlines the concept of an intelligent automotive rear light which is controlled through multiple micro-controllers. The system takes into account the various external influences on an automobile and applies these to the operation of the rear light. Newly developed sensor technologies integrated in advanced micro-systems are capable of providing external environmental data for automatic brightness control within a desired range of light output for constant perceptibility of the light signal to the following traffic.

This new concept will have a significant safety benefit through adaptive perceptibility control resulting in a reduction in road traffic accidents.

CAN field bus technology will be used for communication between the rear lights and other CAN nodes. Algorithms will be used to acquire and provide existing vehicle sensor data to a fuzzy logic controller. An overview of the basic fuzzy control mechanisms for such a system will be given. Various possible micro-system sensors to detect external environmental conditions will be outlined. Safety strategies will be employed to ensure failsafe operation in the case of ambiguous situations or in the event of any faults.

Problem

The following German newspaper article (Sonntag Aktuell, 27.09.98) describes a severe mass accident on the motorway A8. More than 80 vehicles have been involved, three people lost their lives, 47 have been hurt. At the time of the accident, thick fog obviously reduced the visibility by a great extend.

Massenkarambolage im Nebel

Drei Tote und 47 Verletzte bei Massenunfällen nahe Ulm

Ulm (AP) - Bei zwei Massenunfällen auf der Autobahn A8 nahe Ulm sind am

Samstag morgen drei Menschen ums Leben gekommen und 47 weitere zum Teil schwer verletzt worden. Nach Angaben der Autobahnpolizei Mühlhausen waren insgesamt mehr als 80 Fahrzeuge in beiden Richtungen der Autobahn Stuttgart – München in die Karambolagen nahe der Anschlußstelle Merklingen verwickelt, darunter auch zwei Omnibusse.

Zunächst raste kurz nach 08.00 Uhr in Richtung München bei dichtem Nebel ein Bus in ein Auto. Bei diesem ersten Unfall starben zwei Insassen des PKW. In diesen Unfall fuhr ein weiterer Bus und rammte ebenfalls ein Auto. Auch dabei kam dem Sprecher zufolge ein Mensch ums Leben. Aus diesen Unfällen entwickelte sich eine Massenkarambolage, an der letztlich 56 Fahrzeuge beteiligt waren und 39 Menschen verletzt wurden.

Offenbar durch Schaulustige ereigneten sich dann weitere Unfälle in der Gegenrichtung. Auch hier kam es zu einer Karambolage mit etwa 30 Autos. Dabei wurden dem Sprecher zufolge acht weitere Menschen verletzt. Die Autobahn mußte in beide Richtungen voll gesperrt werden. Bis zum Mittag gab es in Richtung München einen acht Kilometer langen Stau, in Richtung Stuttgart waren es rund fünf Kilometer.

Ein Sprecher der Autobahnpolizei bestätigte, daß die seit fünf Jahren betriebene Nebelwarnanlage seit etwa zwei bis drei Wochen wegen eines Softwarewechsels außer Betrieb sei. Seit Einrichtung der Anlage im Jahr 1993 hatte es keine großen Unfälle wegen Nebels mehr gegeben. (Text: AP, Bild: DPA, used by permission)

The previous scenery effected by fog condition shows that environmental effects play a major role in today's traffic. Table 1 shows that about 1/3rd of all accidents [1] go back to darkness and poor road conditions.

Road traffic accidents 1995	Light condition		Road condition	
	Day	Twilight & night	Dry	Other (wet, icy, etc.)
Persons killed or injured	70,1%	29,9%	67,2%	32,8%

Table 1: Influence of Light- and road condition on road traffic accidents 1995

A field study [2][3] has shown that wet weather conditions (e.g. spray and fog) reduces the normal visibility to traffic participants in the front of a driver. On the other hand side the perceptibility of the vehicle itself is reduced as well from the viewing point of the traffic behind. It has been shown before that a vehicle's rear light form the only orientation point in traffic at fog conditions. The reduction of the of spatial perception to a single point – a vehicle's rear light – at poor weather conditions can probably explain, why traffic moves faster compared to the normal.

Coding the braking signal and other rear-light functions via the INTENSITY [4] appears to be highly efficient because an increase in intensity is perceived as a decrease in distance. This regularity in the human spatial perception can be used in supporting the distance regulation. This is especially important for situations with decreased visibility because Cavallo [5] and others have shown that – for instance – in fog distances are underestimated by a factor of up to two. Furthermore, Sivac et al (1987) were able to demonstrate that in this modality the latencies in reacting on the change in a signal are minimal.

Therefore the IntACT project aims to provide a new technology that allows a far more safer road travel under various environmental conditions.

Solution – Concept

Currently the rear light of a car has little intelligence other than it is capable of informing the driver when a bulb is blown it takes no account of weather conditions or the condition of the reflector surface (e.g. amount of dirt) all of which can significantly decrease the amount of light visible by the vehicle behind.

Communicating via the latest CAN environment existing sensor technologies are bridged together with necessary micro-systems control devices (e.g. MOS-FET, power LED's, controller and software) to take account of environmental attributes.

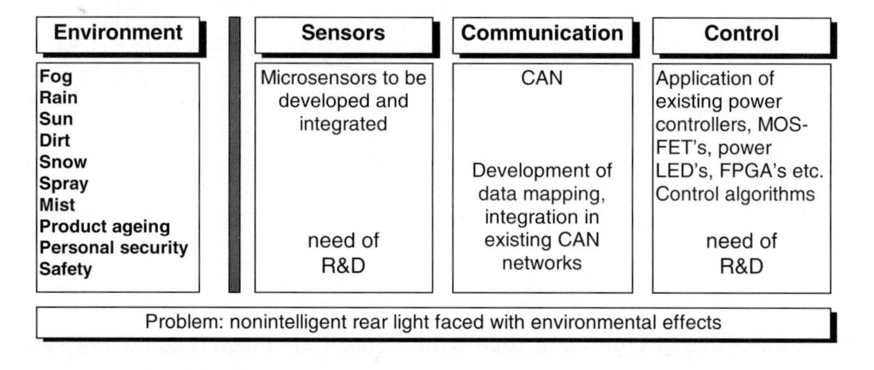

Fig. 1: Environmental influences and key areas for IntACT development

To develop a proper solution, we analysed the light path between the light source (LEDs or filament bulbs) and the observer (in this case the following driver). The produced light output has to pass through several environmental zones which significantly decrease the brightness.

To avoid misunderstandings, the following technical terms were defined and used:

- Transmission: the transmission rate of the outer lens depending on dirt contamination and surface quality
- Visibility: the entirely environmental effects, as there are dust, spray, rain, snow, fog and day / night condition
- Distance: the distance between light source (rear lamp) and observer
- Perception: the "visibility" of the rear light after being influenced by environmental effects over a certain distance and depending on outer lens transmission

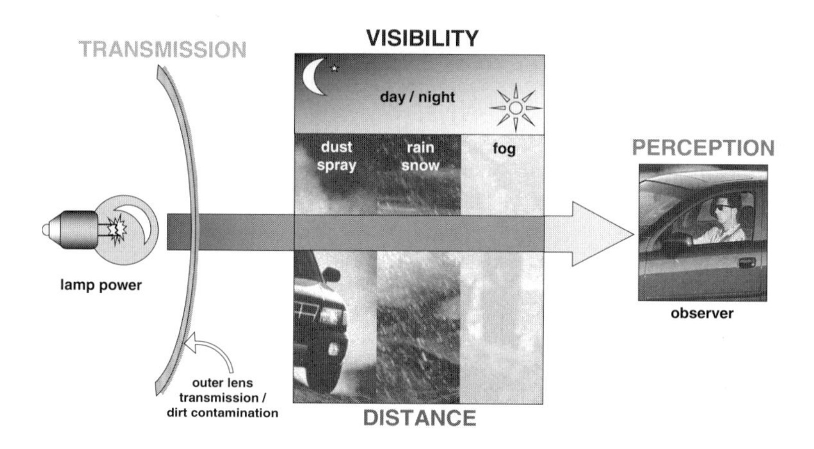

Fig. 2: Environmental influences on light transmission

The goal of the developments is to compensate the above mentioned environmental effects as efficiently as possible by controlling the light output as shown in the following figure.

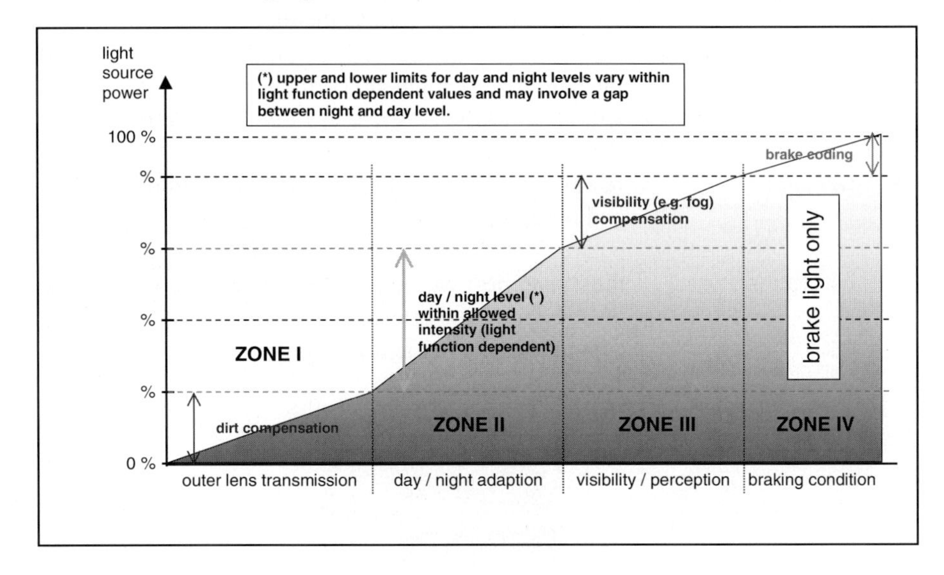

Novel Micro-System Sensors

Visibility and distance sensor

Visibility and distance measurement are realised by the same optical and mechanical hardware.

A first Prototype is implemented with the following geometric-optical dimensions:

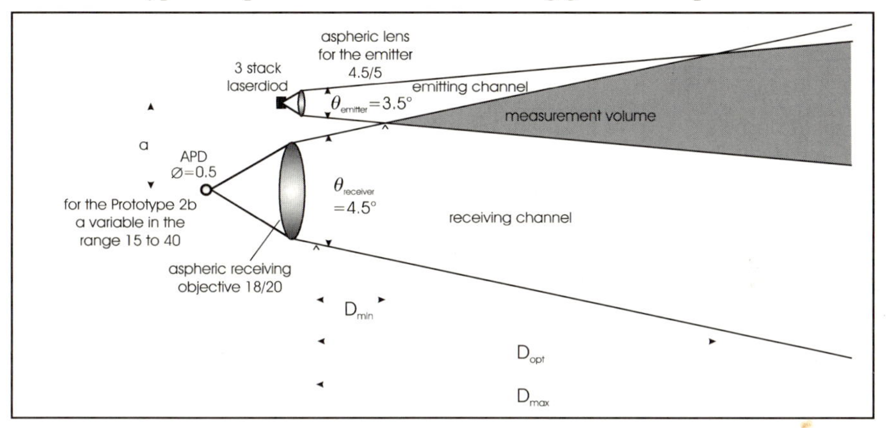

Fig. 4: Visibility and distance sensor prototype dimensions

The maximal achievable distance is shown in the following table :

fog and dust	50...60
road and traffic signs	90...130

The diameter of the measuring volume

$$\varnothing = f(D) = 2 \cdot D \cdot \tan\left(\frac{\theta_{emitter}}{2}\right) + \varnothing_{emitter\,lens}$$

Some samples for the dimension of the diameter are shown in the next table :

D [m]	➦ [cm]
0.7	5
10	53
30	184
50	306
100	612

Surface dirt sensor

Function principle of the sensor

The surface dirt sensor includes two optical principles:

1. Measuring the modulated amplitude of a collimated LED beam, transmitted two times the sensor surface

2. Measuring the modulated amplitude of a diffuse reflected LED beam by the sensor surface

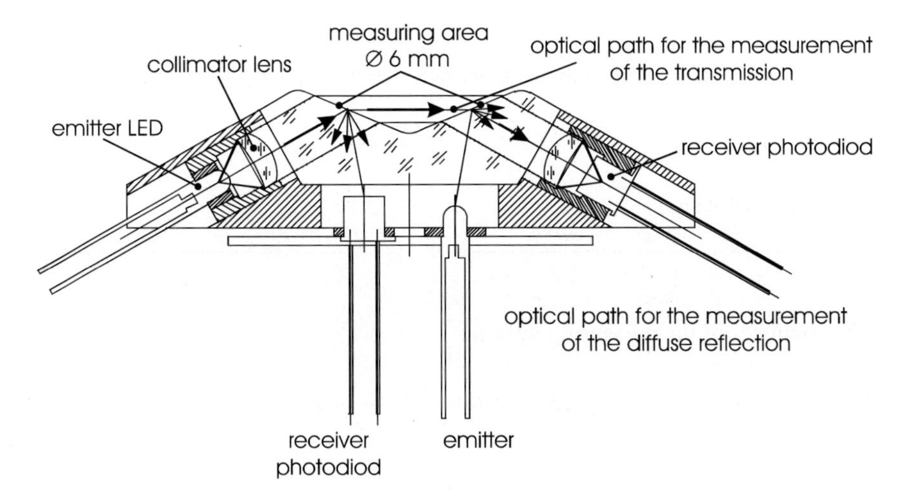

Fig. 5: Dirt sensor concept

The microcontroller compares the signals, generated by the different optical paths and calculates the actual contamination of the surface.

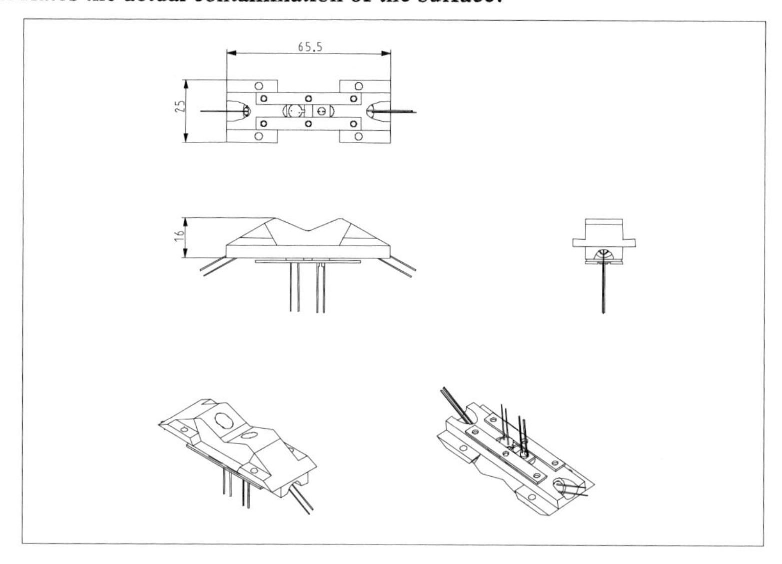

Fig. 6: Geometrical dimension of the dirt sensor

Sun sensor

The actual bright of the sun is measured by the sun sensor.

A large area photodiode detects the quantity of sun light, which is equalised on the whole diode area by a scatter plate.

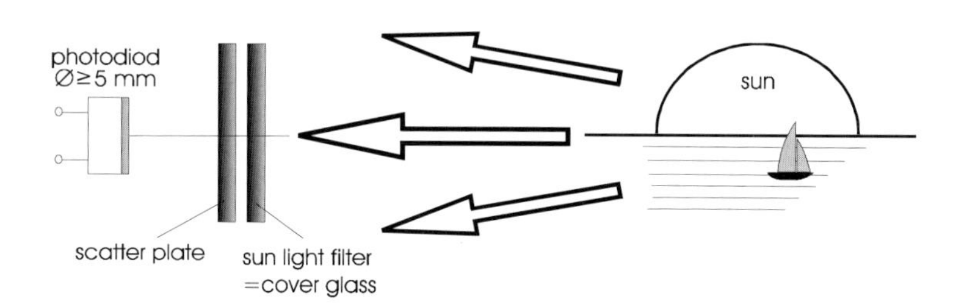

Fig. 7: Sensing of sun light

Complexion and main dimensions of the first sensor system Prototype

The presented single sensors are integrated into a multiple-sensor system:

- The sensor package is assembled in a waterproof plastic case,
- it is covered by a red cover glass which has a filter effect.
- The mounting area of the sensor package is situated at the rear side of the case.
- It is planed to install the Prototype 2b outside the car. For first tests under real conditions, the sensor will be mounted near the licence plate on the rear side of a car.

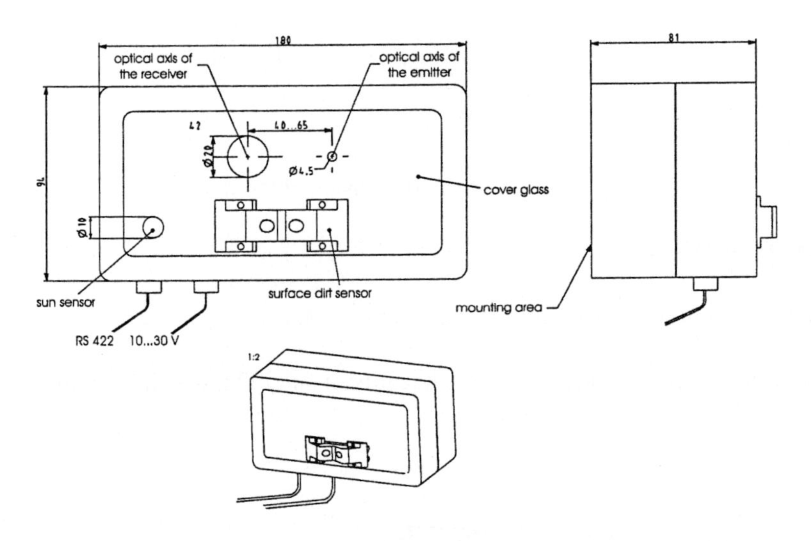

Fig. 8: Prototype of combined sensor system

Algorithms & Control

Fuzzy Weather Model Controller

It is the function of the Weather Model Controller to determine a value of perceptibility, p, from the four sensor inputs: visibility, dirt, range and ambient light.

The fuzzy modelling approach was chosen, as it has proved to be suited to problems in which the modelled relationships cannot be easily determined, are highly mathematically complex, and/or where the inputs introduced are inaccurate. This approach is based around the descriptive terminology a human would use when implementing control of this type, for example:

If visibility is *poor*, increase the rear light intensity *quite a lot*.

Systems that can be controlled by humans in this manner, are by their very nature suited to fuzzy logic control.

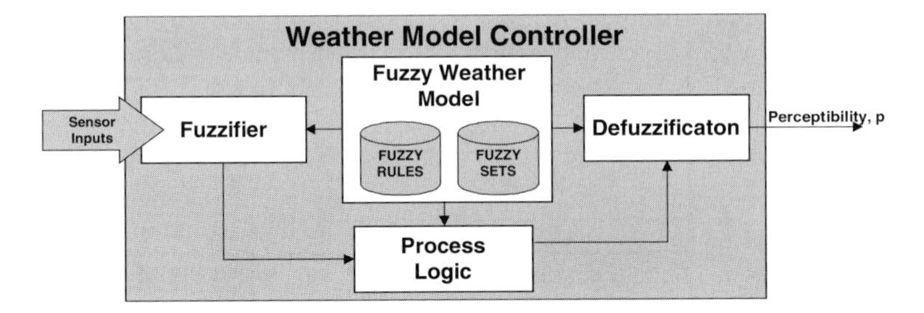

Fig. 9: Weather model controller

Figure 9 shows a fuzzy logic rear light controller. As shown this comprises four main sections: the fuzzifier, the process logic, defuzzification and the fuzzy weather model. It is the last of these that provides the information required to convert the sensor input information into an output value of perceptibility, p, which is used to control the brightness of the rear light. The steps involved in obtaining this value, p, from the different sensor inputs is outlined below.

Fuzzification Module

For each of the measured input sensor variables, Q_X (where X = V, D, R, A, representing visibility, dirt on rear lens, range from the following vehicle and ambient light, respectively), there is a set of fuzzy control regions, known as the term set. Figure 10 shows the fuzzy set regions describing the measured visibility input, Q_V. As shown, the visibility conditions are described by the fuzzy regions: Very Good, Good, Medium, Poor and Very Poor.

A particular value of measured visibility, Q_V, will have membership of one or two of these fuzzy regions. This is described by *the Degree of Membership*, μ[x], which is normally a value from 0 to 1. It is the values of these degrees of membership that form the output from the fuzzification module.

For example, Figure 10 shows how for a measured visibility value of Q_V:

> the degree of membership, μ[v], to the region *Poor* visibility = 0.6;
>
> and the degree of membership, μ[v], to the region *Medium* visibility = 0.2.

Hence it could be said that for a measured value of visibility, Q_V, visibility may be described as medium-poor.

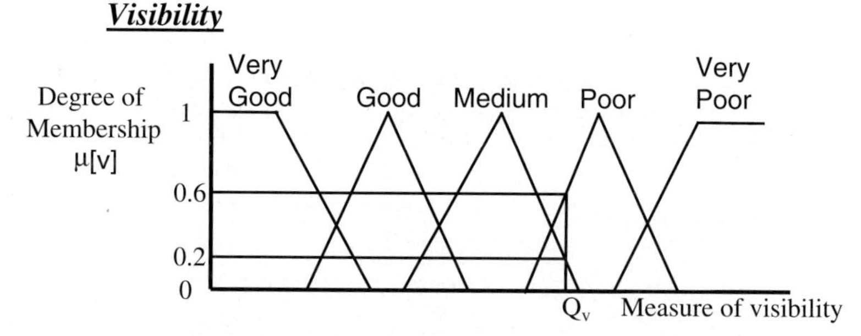

Fig. 10: Visibility fuzzyfication

Process Logic Module

The values of μ[x] determined in the *fuzzifier* block of the fuzzy controller shown in Figure 9, are fed into the *process logic* block. It is here that the type of control action, although not yet the degree of action, to be taken is determined from a set of rules (stored in the Fuzzy Weather Model). These rules are of the form shown below:

If Ambient_Light is *Dark* and Range is *Close* and Visibility is *medium* and Dirt is low

then Perceptibility is *QUITE HIGH*

If Ambient_Light is *Dark* and Range is *Medium* and Visibility is *poor* and Dirt is medium

then Perceptibility is *LOW*

As shown, a combination of fuzzy inputs (i.e. ambient light, range, visibility and dirt) determines a fuzzy output for perceptibility - for which there is an additional fuzzy set.

For the four measured sensor inputs, one or more rules may be true, to differing degrees. The membership of each of these rules, are determined from the degree of membership to the appropriate regions within each term sets, i.e.

$\mu[a], \mu[r], \mu[v], \mu[d].$

Which are fired in parallel and used as inputs to the relevant regions (based on the rule set) within the perceptibility term set.

Defuzzification Module

Defuzzification is the final step involved in the Weather Model Controller. It is here that the fuzzy perceptibility inputs, obtained in the *process logic* module, are used to determine a crisp output of perceptibility, p. This may then be used to determine the brightness levels of the rear light.

Intelligent Power Control

The Intelligent Power Controller controls the output of the rear light, in terms of brightness control and LED functionality selection (as described in the following sections). The inputs to this controller include:

the value of perceptibility (obtained from the Fuzzy Weather Model Controller),

rules based on the legal brightness levels for road vehicles and preferred LED cluster functionality (which may vary depending upon the road conditions and in the event of LED or light failure).

LED failure information – obtained by the controller itself (as described below).

Brightness control

PWM was found to be the most efficient way of controlling the brightness of the rear lights. The main advantage of a PWM system initially is it's efficiency which is at present in excess of 95%. The frequency of the PWM system is set to 200Hz.

The brightness of the rear light is adjusted by varying the level of PWM. Initially a fixed PWM level is set for each different rear light function (i.e. fog, tail, stop and turn). Changing environmental conditions will then lead to changes in the PWM levels.

LED failure detection

Using sense resistors the intelligent power module also has a built in analogue LED failure detect system, informing the driver of the failure and re-allocating unused LED's to make up any deficit. This system has also being successful in detecting any potentially hazardous LED conditions e.g. over temperature condition.

LED functionality selection

The LEDs are selected for PWM control in groups of 3 or 4, allowing different groups to be selected in the event or LED failure. This will also allow more perceptible rear light functions e.g. the whole rear light could be activated in the event of emergency braking.

Other functions

Another aspect of the intelligent rear light power control is that it can be used to implement safer braking patterns. BMW produced a paper [6] that highlighted the need for a more visual display of braking strength. Based on their own research they came up with seven possible brake encoding forms. Their results for an optimised display of braking force is a three stage display, with combined change in surface area, position and luminance. The intelligent power control can implement all these functions.

Using an intelligent control PWM system also allows the possibility of car to car communication. In principle the technique would work by leaving the PWM value constant and varying the PWM frequency, i.e. a car to car FM communication system using the rear light. An FM system also ensures that the perceptibility of the lights remains unaltered while in communication. However care in the design would be needed to avoid unwanted crosstalk between vehicles.

Intelligence needs communication

The coherence between the ability of communication and the efficiency of a system gets more and more important in modern motor vehicles. The number of electronic control units (ECU) increases rapidly. Also the performance and capabilities of such systems increase very fast. The connection of several ECUs to a serial network allows the implementation of an intelligent car system instead of using only stand-alone devices. Today most European car manufacturer are using the Controller Area Network (CAN) as communication medium for data transfer between the several ECUs.

Overview on CAN-based higher layer automotive protocols

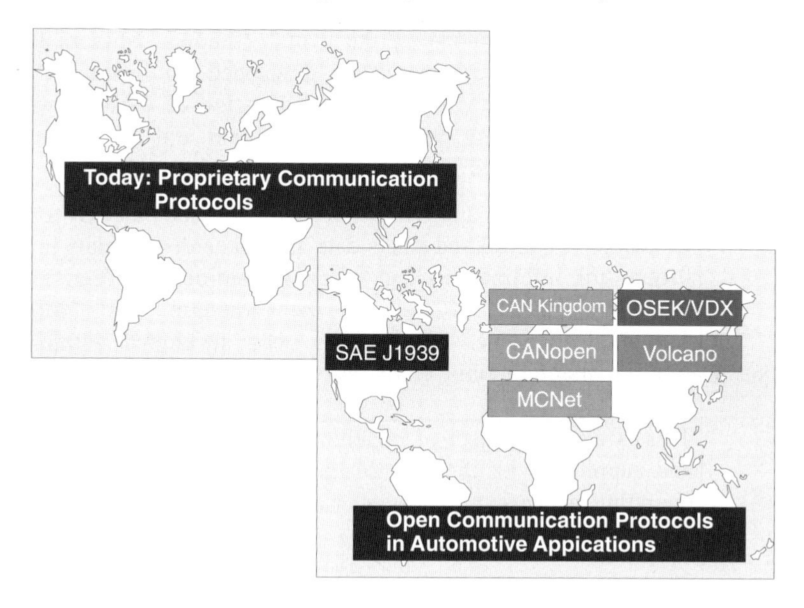

Fig. 11: The world of automotive communication protocols

This communication and coding rules are called language or in a technical background communication protocol.

Within the scope of the IntACT development project, the variety of CAN-based communication protocols has been considered. Further the several communication protocols has been compared and assessed. The point of view was the usability of the protocol for the communication between both rear light control units.

This investigation shows on the one hand that today most car manufacturers are using proprietary communication protocols in their vehicles. But on the other hand there are several open communication protocols available for use in automotive applications. Figure 11 illustrates this situation in the world of in-vehicle communication protocols.

In the IntACT development project following open communication protocols have been considered.

SAE J1939	This protocol is specified by SAE (Society of Automotive Engineers). It is based on CAN specification 2.0B (29bit identifier). J1939 originally represents a protocol specification for heavy trucks and busses. However, it is more and more used also in other kinds of motor vehicles.
VOLCANO	This is an open protocol of the British company Nothern Real-time Technologies. It has been developed on behalf of two Swedish car manufacturers. For the creation of CAN networks under Volcano, very powerful development and configurations tools have to be used.
MCNet	MCNet is an open communication protocol based on CAN for the exchange of control and status data as well as display data in text format for information- and entertainment-devices (e.g. radio, CD-player etc.). MCNet is the acronym for "Mobile Communication Network". It has been developed within the Bosch division "mobile communication".
OSEK/ VDX	This is a joint project of different European car manufacturers and suppliers with the aim to develop standards for interfaces, operating systems and protocols for all electronic control units in cars.
CAN Kingdom	This protocol has been developed by the Swedish company Kvaser in co-operation with further manufacturers of hydraulic heavy machines, like for example harvesting machines. CAN Kingdom provides no ready-to-use communication software but a library of protocol-primitives.
CANopen	CANopen has been defined in the first place for use in automation systems. But due to its general purpose features, it is already gaining influence in non-industrial sectors including some special automotive applications, like for e.g. fork-lifts.

Table 2: Open automotive communication protocols

Communication data overview

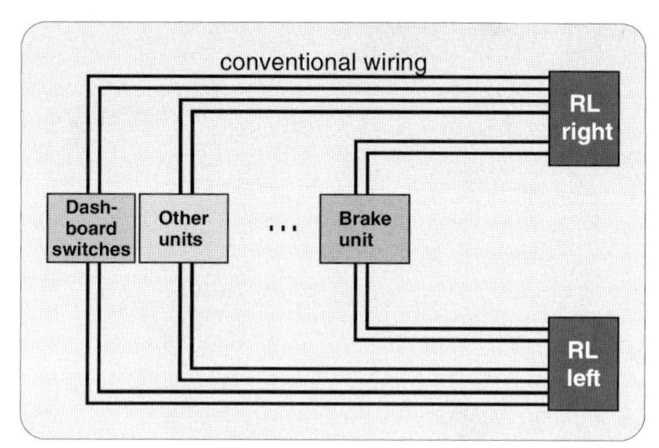

Fig. 12: Elementary data connection

On step in the development of an intelligent rear light is to analyse the different kinds of data transfer between the rear light and the other ECUs in the vehicle. Therefore it is necessary to classify the different information and to structure this data in the following types:

- "**Fundamental data**"

  Today the rear light is not intelligent. The light sources are only switched by several units of the car (see Figure 12). In the most cars of today the switches are connected to the rear light by conventional wiring.

  This means that each information (e.g. direction status, brake signal, etc.) is a one bit digital data (light on/off).

  This elementary data must be made available to the rear light every time. This is an absolutely strict requirement for all car communication systems and is also an elementary function of an intelligent rear light system.

  This fact is the base of all further development steps within the IntACT project. The next figure shows all types of information which can be important for a intelligent rear light controlling system together with possible interfaces.

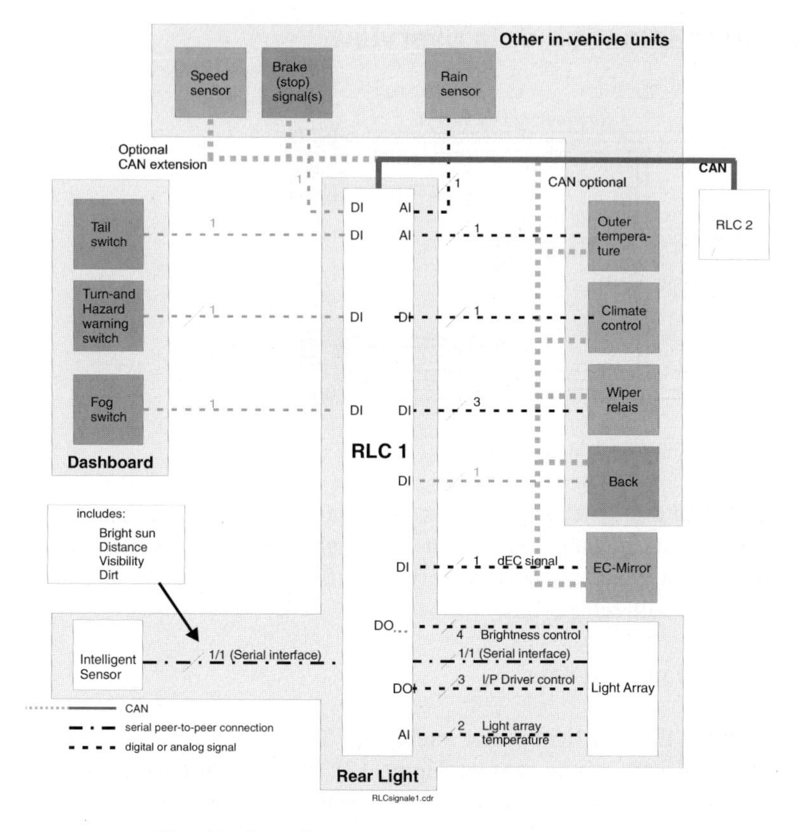

Fig. 13: Rear light control data overview

Beyond this it is necessary and important to get access to the following types of data:

- **"Nice to have"-data**

  This are additional data for e.g. environmental data which are important to get a more exactly brightness adaptation of the rear light. The number of provided signals are different in each type of car. This signal are provided via conventional wiring or via CAN connection.

- **Rear light controller information data**

  The intelligent rear light system is not only a data receiver. Each rear light controller also provides information to other ECUs and to the second rear light controller. These are data from the intelligent environmental sensor together with status information (e.g. actual brightness level, error messages, etc.) of the rear light controller. This data are used to have a plausibility control between both rear light controllers and to notify error states.

Step-by-step-integration in existing car CAN communication concepts

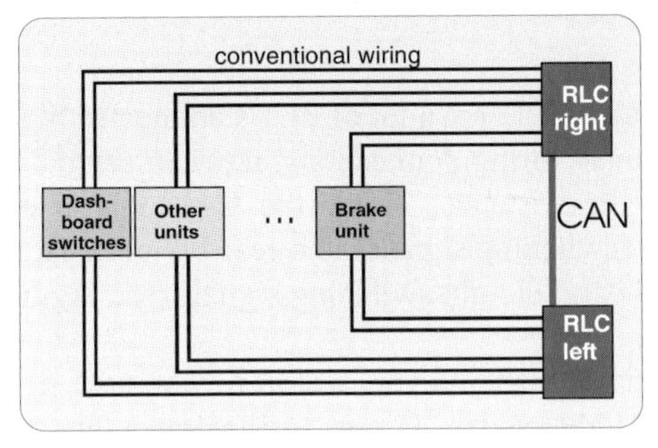

Fig. 14: Basic CAN communication

On the way to integrate the intelligent rear light system into a series car completely, it is necessary to define clear integration steps. One of the key function in this way is the step-by-step increasing of the CAN communication.

CAN communication between rear lights

With this background, the minimum requirement for the IntACT communication beside the reaction on the elementary data information is a CAN communication between both rear light controllers. This is the first and easiest step, because the CAN communication does not have any influence to other part of the in-vehicle CAN communication. In this first step CAN offers an easy way to exchange information of the intelligent sensors, values of the brightness control software, status information and emergency messages. Figure 14 shows this basic CAN communication version of an intelligent rear light system in a car.

Rear light controller connected to body-electronic-CAN as passive data receiver

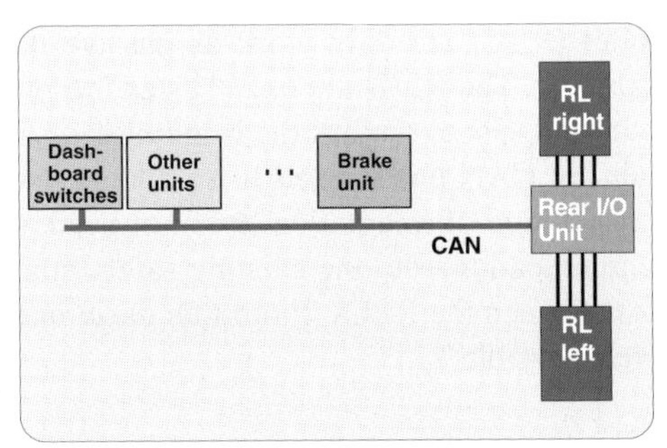

Fig. 15: CAN in today's upper class cars

The absolutely basic communication via CAN in an intelligent rear light system (see Figure 14) is the starting point of the next steps of full integration in modern series cars.

Today in upper class cars there are a lot of ECUs connected together via CAN bus. The former conventional wiring connections between several ECUs and the rear lights are not longer in existence. Today the CAN bus is used to transfer the information (e.g. light switching signals) to a rear I/O unit. This ECU transfers the serial bus signals into parallel light switching signals (see Figure15).

This kind of information transfer is very useful for the next step of the integration of the intelligent rear light system. Figure 16 illustrates this next version of a car with CAN connected intelligent rear light system.

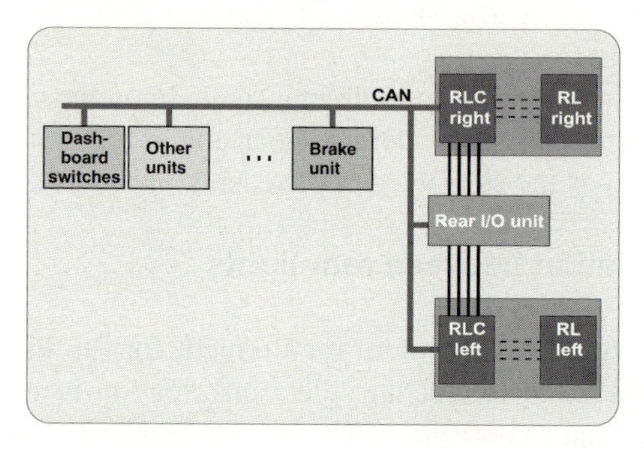

Fig. 16: Integration in upper class cars

The greatest advantage of this integration step is that the integration of intelligent rear lights in cars can be done without major modifications of existing communication functions. In this version of integration the safety relevant functions ("fundamental data") are transferred in the same way as in today's upper class cars.

The rear light controller (RLC) are only work as passive data receivers on the existing CAN bus of the car. The RLCs are only "listen" to messages for other in-vehicle ECUs which are relevant to control the brightness of the rear lights ("Nice to have"-data). Further the CAN bus is used to exchange data between the both rear light controllers. This kind of integration does not need any intervention into the existing CAN communication, into the existing safety strategies and gives the feature of data transfer between the RLCs. From this basic integration concept are further integration levels possible.

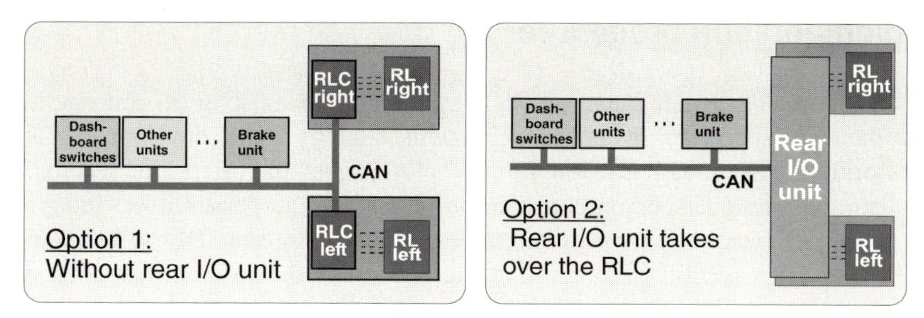

Fig. 17: Further integration possibilities

It will be possible in future that the RLCs takes over the functions and tasks of the rear I/O unit. A other possibility of integration is that all functionality of the intelligent rear light controllers (e.g. sensor data processing, brightness control, etc.) are implemented into a powerful rear I/O unit.

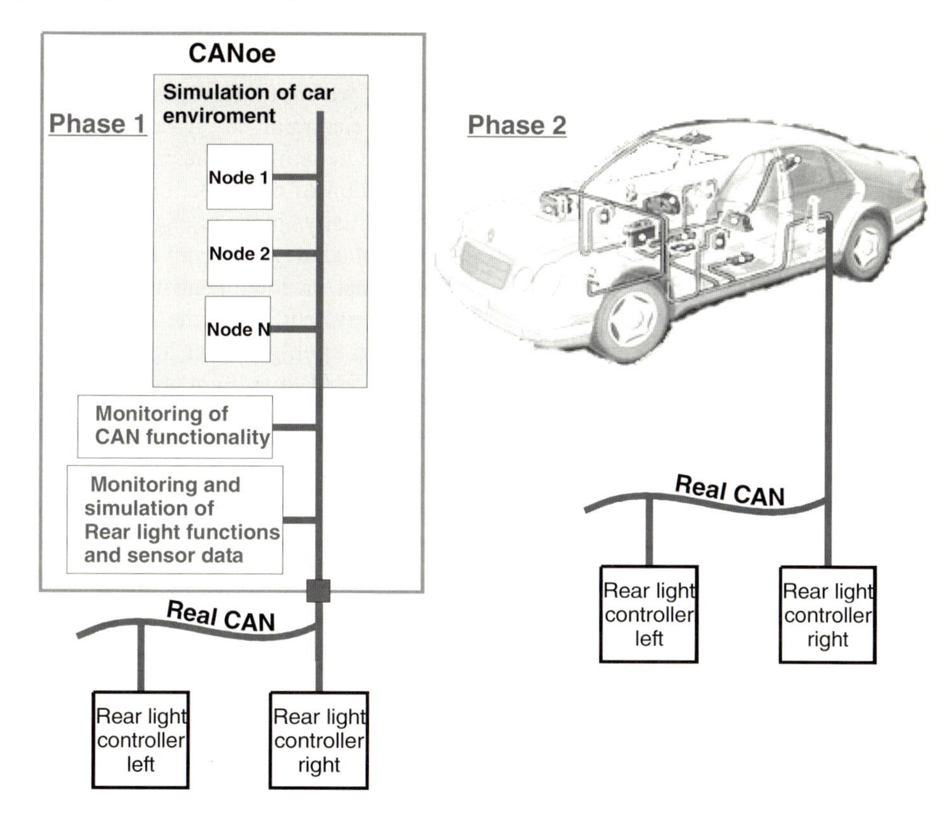

Fig. 18: Implementation steps

Implementation procedure

For integration of the intelligent rear light system into an existing car concept, it is important to apply an overall development strategy. The development and monitoring software-tool CANoe from Vector-Informatik offers the feature to simulate the whole CAN communication of a car and the possibility to integrate the rear light system step by step into the existing car concept. This software tool offers also the possibility of monitoring and displaying the different functions and values of the rear light system. The way from simulation (CANoe) to integration of real hardware and software is shown in Figure 18.

Increase of safety by intelligent rear lighting

An intelligent rear light system together with an integrated communication concept provides the following aspects of increased traffic and driver safety:

- **"Plug&Play&Listen–concept" – Safety without interference to existing car communication concepts** Today, the electronic car systems are modular and intelligent by themselves. By in-vehicle networking, the car becomes more or less a homogenous system. The integration of new electronic systems in a given communication concept causes usually high re-engineering effort. According to this fact, the IntACT rear light concept will prove a possibility to exchange the conventional real right against an intelligent rear light system without affecting the total communication system. After the exchange of the rear light all additional safety features of an intelligent rear light are available ("Plug&Play"). Therefore the CAN-connected rear light system works as passive listener to get additional information for controlling the rear lights.

- **Self controlling system by continuous data exchange and data adjustment** The normal road traffic causes environmental conditions, which are only partial present at one of both rear lights. Examples are a motorbike behind one rear light or shop window lights on the roadside. These partial influences have to be balanced for controlling the average brightness of the rear lights. Therefore, the IntACT rear light system requires cyclic data exchange and adjustment via CAN between both lights. Additionally, the implemented communication allows plausibility checks and in due course the detection of emergency conditions in system components.

- **Adaptive redundancy – Automatic compensation of failed light sources** Rear lights are subdivided in different fixed function sectors (e.g. tail light, fog light, etc.). At the breakdown (e.g. light source failure) of one light section, the function is usually completely out of order or at least strongly reduced in brightness. This is a high safety risk for all traffic participants. The IntACT rear light allows e.g. the logical replacement of failed light sections by others.

- **Intelligent light controlling offers higher traffic safety** The safety level of present rear lights is drastically reduced in the case of bad weather conditions (e.g. fog, rain, etc.) and also by wrong driver handling (e.g. no foggy conditions, but the fog light is switched on). The present safety level is only the base level for an intelligent rear light system. The high integrated sensor module for detection of safety relevant environmental conditions and the continuous balance with other in-vehicle data sources via CAN network, together with fuzzy-controlled brightness and the using of modern light sources and new electric connection systems offer a strong increase of active traffic safety. By use of an intelligent rear light system, the goal is to minimise the influence of weather conditions on the visible brightness of the rear light. The brightness level will be constant in the view of a following traffic participant. In the future, the function level of a present rear light (switch on/off) is only the last fall-back level at the breakdown of all microcontrolled electronic rear light functions. Besides the brightness control, an IntACT rear light system can offer additional functions (e.g. automatic light flashing in the case of panic braking) to get higher active traffic safety, too.

Prototype-Integration

IntACT aims to use multiple sensors, advanced algorithms and CAN communication and control techniques. The major innovations of IntACT are measurable at two levels of prototypes (A and B) that are enriched with miniaturised macro technologies:

- Reliable measurements of environmental effects through multiple sensors combined with advanced algorithms

- LED power control through MOS FET technology

- Ability of communication of a rear light (integral part)

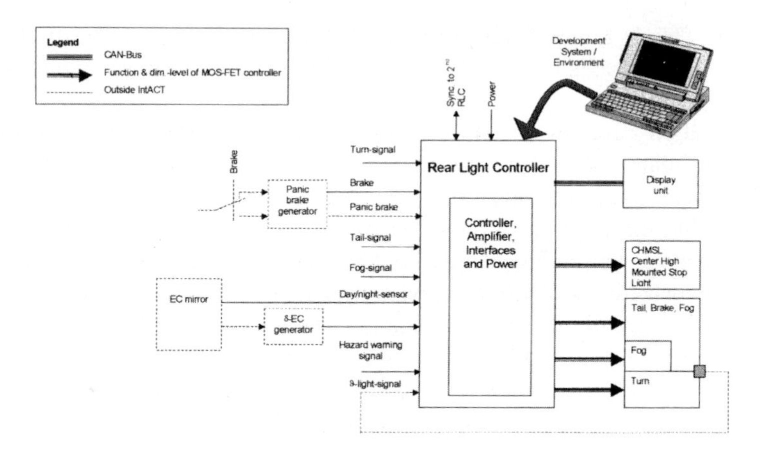

Fig. 19: Prototype A, schematic diagram

The first light unit (prototype A) requires the application of microprocessor control and fault tolerance intelligence in terms of power control (with daylight sensing light control) whilst the second (prototype B) has to incorporate all of the sensor miniaturisation, CAN communication and weather control algorithms.

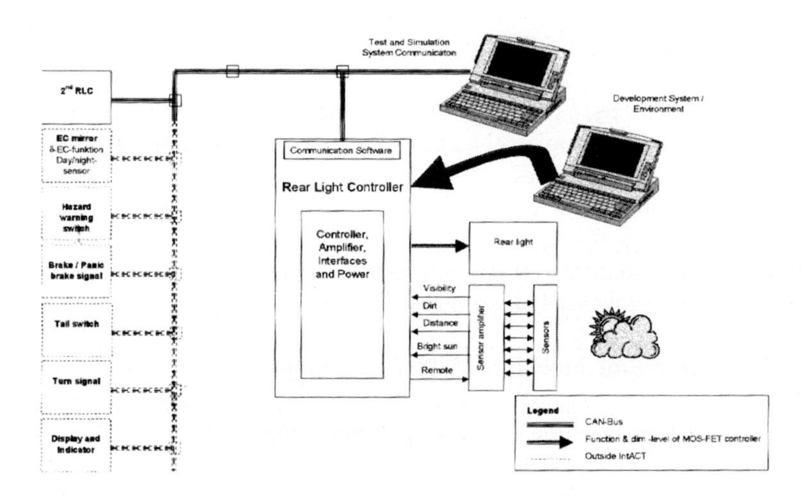

Fig. 20: Prototype B, schematic diagram

There is a market need for both types of rear lights because the first is lower in cost and can therefore be sold in lower priced cars to a larger market.

Currently prototype A is under development.

Conclusion & Future Aspects

Tomorrow's cars will be equipped with a lot of additional sensor information and security devices compared to today's vehicles. Mobile measuring systems installed in every single car will take the place of local and road based information systems (e.g. fog measuring system at autobahn A8). Even in today's and near future's upper class cars such devices have been integrated. The new Mercedes S-Class is optionally equipped with a radar based forward distance sensor and the soon coming BMW 7 seems to be provided with a three level stop light, indicating braking force by increasing the stop light area.

Recently there is a study in progress about the social benefits of the technology within the EC in terms of accident reductions, reduced hospital costs, lost working time and reduction in traffic congestion due to reduced accidents.

References

[1] Statistisches Bundesamt: Verkehrsunfälle 1995, Fachserie 8, Reihe 7, Wiesbaden 1996

[2] Prof. Steinbach, Peter: "Massenunfälle bei Nebel", Berichte aus der Arbeitseinheit Sozialpsychologie, Ruhr-Universität Bochum, Fakultät für Psychologie, 1995

[3] Koenig, Dr. Josef, "Massenunfälle im Nebel - Preiswürdig", 2. Fritz Thyssen Preis für Bochumer Sozialpsychologen,

http://idw.tu-clausthal.de/archiv/all/mail.893767043.18474.html

[4] Zimmer, Fortschritte der Verkehrspsychologie, Kongressbericht, Bernhard Schlag (Hrsg.); Deutscher Psychologen-Verlag, 1997

[5] Cavallo, Viola, "Distance estimation in fog and fog light design", Symposium "Perceivable Danger", University of Regensburg, 1997

[6] Fenk, Johannes, "Efficiency of a braking intensity indicator", BMW AG

Miniaturized Scanning Laser Radar for Automotive Applications

M. Monti, K. Bettaieb, L. Zago, P. Debergh, Y. Welte, Y. Depeursinge

CSEM SA - Centre Suisse d'Electronique et Microtechnique
Jaquet-Droz 1, CH-2000 Neuchâtel, Switzerland

Abstract: In the frame of an ESPRIT project called OLMO, CSEM has developed an innovative pre-industrial scanning range sensor based on laser radar techniques. OLMO is the acronym for "On-vehicle Laser Microsystem for Obstacle detection", and the project was carried out in collaboration with Fiat Research Center (I), Magneti Marelli (I), Renault (F), CEA-LETI (F) and Jenoptik (D). Major results of this project are presented as well as further developments.
The key-features of the system are high compactness (less than a packet of cigarettes), low cost and high performance in adverse visibility. The system is based on the measurement of light time-of-flight by the means of an innovative signal processing algorithm and on a resonant, frictionless mechanical structure providing a sinusoidal horizontal scanning both of the emitted laser beam and of the detection optics. The signal processing technique, called Random-Modulation time-of-flight (RM-TOF), is inspired by the spread-spectrum techniques used in the field of telecom and GPS, and exploits the auto-correlation characteristic of pseudo-random sequences. Major advantages of this technique are robustness and the possibility of using a commercially available low power laser diode source (20 mW), thus reducing the costs. The scanning mechanical structure, which was inspired by mechanisms used in aerospace applications, consists of a resonant structure based on a *Flextec* pivot element composed of two cross-flexure elastic structures obtained by layering. Flexure structures are frictionless components in which kinematic joints are materialized by elastic thin beams. Major advantages of this concept are very high compactness, very long lifetime, low cost and low power consumption.

Keywords: Laser Radar, Range Sensor, Scanners, Random-Modulation Time-of-Flight, Obstacle Detection, ACC.

1. Introduction

In November 1996, an ESPRIT project was started, called OLMO (On-vehicle Laser Microsystem for Obstacle detection). The aim on this project was the development and the on-road testing of a pre-industrial scanning range sensor

based on a 1.5 μm laser technology, which should be extremely compact and low cost, while allowing high performances both in good and in adverse visibility. The partner were Fiat Research Center (I), Magneti Marelli (I), Renault (F), CEA-LETI (F), Jenoptik (D), and CSEM (CH).
During this project, two prototypes have been developed for testing two different approaches. Both prototypes use the same scanner (CSEM) described hereafter, but one is based on a pulsed microlaser chip (LETI and Jenoptik) and a Time-of-flight signal processing (LETI), and the other is based on a commercial low power cw-laser and a Random Modulation Time-of-flight algorithm (CSEM).
Since the microlaser/TOF approach has already been published and presented at IBEC '97, this paper will focus on the cw laser/RM-TOF approach.

2. System Specifications

The system is based on the measurement of light time-of-flight by the means of an innovative signal processing algorithm and on a resonant, frictionless mechanical structure providing a sinusoidal horizontal scanning both of the emitted laser beam and of the detection optics. The whole system is designed to fit in a very compact package, in order to be easily mounted in the car front panel or inside a headlamp. The target specifications are the following:

eye safety :	class 1
wavelength:	1550nm
range in good visibility:	80m on diffusing target, 200m on retroreflector
range in bad visibility:	40% better than human eye
range accuracy:	0.1m for range <30m, 0.5m for range > 30m
horizontal field of view :	20°
angular resolution:	0.2°
frame rate:	14 to 20 Hz
output:	target distance vs angular position
output rate:	2 kHz

3. Scanning Mechanism

The scanning system allows the angular motion of both the emission and detection light beam over a range up to 20° (+- 10°). The design of this active mechanism is driven by the optical requirement for the detection module, in particular:
- an aperture of the order of 20 to 25 mm diameter,
- an optical design providing the optimal efficiency for the detector.

Other main design drivers are related with the issue of large scale, low cost production:
- compactness and miniaturization. The overall dimensions of the system are a first priority issue, specially concerning width and height (front view).
- minimization of the number and the cost of components, in order to reduce production and assembly costs (use of low cost plastic components).

The emission laser unit is fixed and emits vertically a divergent lightbeam 0.3°x3°. This light-beam is deflected by a mirror mounted on the scanning mobile part, providing therefore a scanning illuminated spot which is 0.3° wide and 3° high in front of the car.
The complete mobile part (including the detection unit and the attached prism) is mounted to a fixed support by flexure pivots. On the fixed part, an electromagnetic drive, a motion detector and a control electronics are located, which allow the activation and the control of the scanning motion.
The element which materializes the pivot consists of two cross-flexure elastic structures obtained by layering (CSEM patent pending). Flexure structures are frictionless components in which kinematic joints are materialized by elastic thin beams. This ensures in particular a long lifetime. The scanning oscillation is driven by the resonance frequency of the flexure pivot with the inertia of its mobile load. The mobile part is balanced about the scanning axis in order to be insensitive to external transverse vibrations.

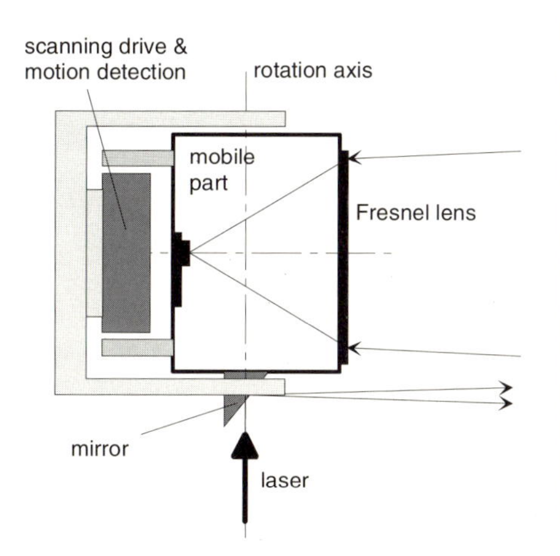

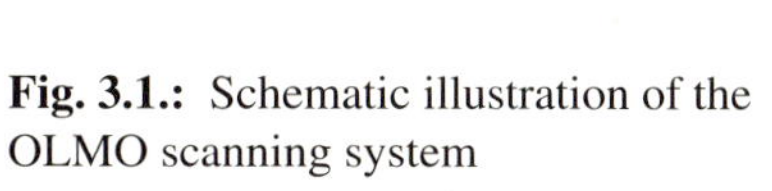
Fig. 3.1.: Schematic illustration of the OLMO scanning system

Fig. 3.2.: CAD picture of the scanning mechanism (back/bottom view).

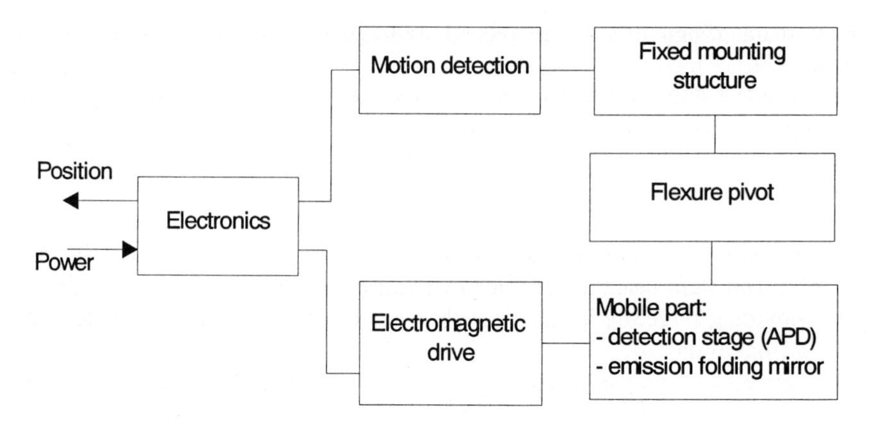

Fig. 3.3: Functional schematic of the OLMO scanning concept

The activation of the scanning system is obtained by means of a small linear electromagnetic drive, essentially constituted by two cylindrical coils as the stators and two iron elements as mobile components. Because the scan motion is performed in a resonance conditions the drive only has to provide the tiny acceleration pulse (2 ms) required to bounce the scanner and keep the desired amplitude of the resonating system. The power consumption is therefore very low. The motion of the scanner is detected by a cheap but efficient optical probe. The signal is then used by an electronic circuit to trigger the motor pulse. Main data of the OLMO scanning system are:

Scanning frequency	7 to 20 Hz
Total scanning angle	max 15 deg
Scanning function	sinusoidal with cuspids
Overall size	30x37x40 mm (wxlxh)
Detection lens aperture diameter	22 mm
Mobile part: mass ; inertia about axis	15 g ; 40 g cm2
Actuation pulse	12 V, 2 ms
Power consumption	< 30 mW

4. Random Modulation Time-of-Flight (RM-TOF)

A distance measurement is initiated with a start signal which causes the emitter to send a train of coded pulses. The echoed signal is detected with an avalanche photo-diode (APD). The current generated by the APD is amplified with an automatically controlled gain and digitized with an ADC for further digital processing.

The distance is measured in a two step process: an acquisition step that gives a coarse measurement and a tracking step that provides a fine measurement. The measured delays are then calibrated with respect to the fixed delays introduced by the detector.

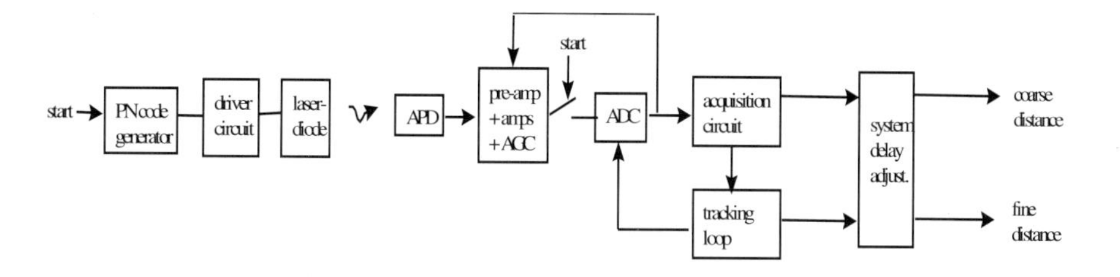

Fig. 4.1. Random modulation time-of-flight system

Acquisition

At a first step, the time-of-flight delay is measured with the acquisition circuit with a rough resolution of one chip period, Tc (i.e. the time to transmit one pulse of the code).

The sampled data is fed through a correlator which uses a replica of the original code to measure the degree of alignment (coincidence) of the received signal and the emitted one (i.e. the output of the correlator is maximum for perfect alignment). The correlator output for different time shifts (i.e. N shifts where N is the code length) are stored and accumulated for different sequences. The maximum corresponds to an alignment of the two codes within one chip period. In order to avoid false alarm, the maximum is valid only above a certain threshold. To achieve a pre-specified false alarm rate, the threshold is set automatically by measuring the noise level. The measured time-of-flight delay is given by nTc where n is the index of the time shift that gives the maximum auto-correlation value.

Implementation

After successful simulations and emulations on real data in different atmospheric conditions (including fog), this algorithm was implemented on a FPGA. The complete HW consists of two printed-circuit boards (PCB), an optical head, a scanner, and a laser diode module. One PCB provides all the supply voltages for the system from a single 12 V supply and the other PCB implements the digital functions. The optical head includes the receiving lens, the APD and the analog front-end which converts the APD signal into a TTL-level digital output. The different interfaces to the digital PCB are shown in the following block diagram.

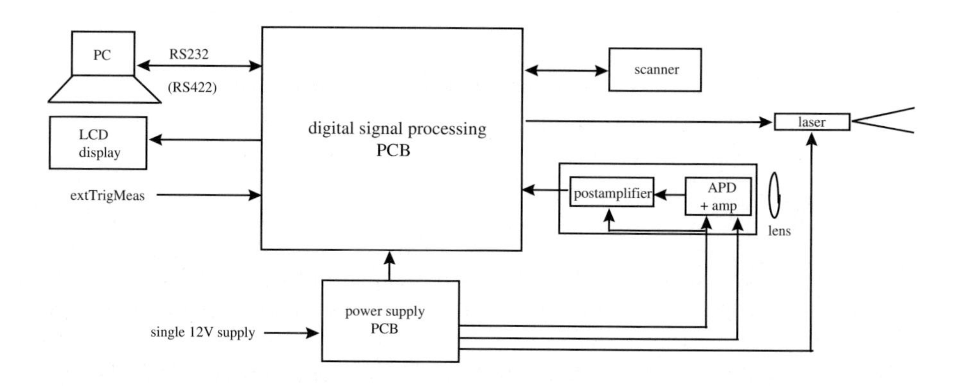

A PC is used to communicate with the digital board either to program some parameters of the application or to receive the measurement data. The latter are also provided in a parallel form that can be used by an LCD display. For each measurement, the scanner is prompted to provide the current angle which is appended to the distance data. The internal clock frequency corresponds to a scanning rate of 20 Hz where a frame is divided into a 100 separate targets. The laser beam is generated by a complete module that includes the laser diode, the driving electronics and peltier-based temperature regulation for wavelength control.

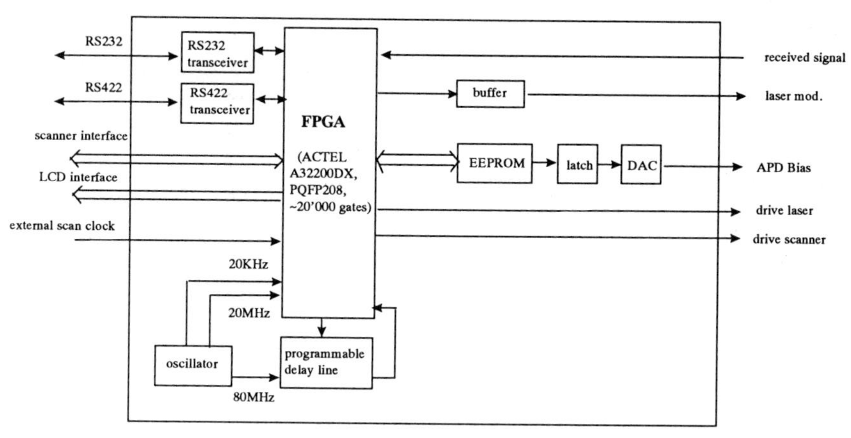

Fig. 4.3: Architecture of the digital PCB

The signal processing functions are performed by a single FPGA. This provides for a robust design with a minimal component count. The functions implemented by the FPGA are : a) RM acquisition circuit; b) RM tracking loop; c) command interface; d) transmission of measurement data.
The command interface allows to control various aspects of the application from a PC. Commands are issued to the PCB through a serial line. Some commands allow the modification of various parameters of the application as well as transfer debugging data from the FPGA to the PC. The parameters are saved upon request to an EEPROM on-board. They are automatically loaded at reset. After each measurement, data is transferred both through the serial line and the parallel LCD interface. The data includes the angle, the coarse and fine distances, the amplitude and some status information.

5. System Integration

A first operational prototype of the complete system has been realized in a self-consistent package. The size of this prototype (see figure 5.1.) is about 4 times (in volume) the final size of the product, which is shown in figure 5.2.
This prototype integrates the complete scanning mechanism (i.e. mechanics, actuation, position measurement, detection optics and detection electronics), power supply (5V, 60V and stabilized 12V), a 20 mW CW laser source, the RM-TOF electronic board, and RS-232, RS-422 and parallel output connectors. It is described in figure 5.1. and 5.2.

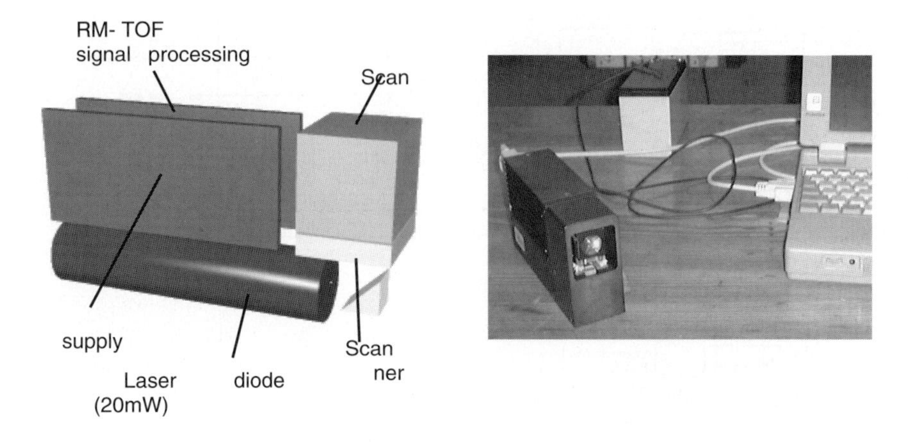

Fig. 5.1: Operational prototype.

Fig. 5.2: Planned final size of the complete system including scanner (emission & detection), laser, signal processing, communication.

8. Conclusions

The scanning mechanism has been tested extensively both in laboratory (CSEM, CRF) and in on-road conditions (CRF), where the expected performances were totally confirmed, meeting the specifications.
The feasibility of the RM-TOF algorithm has been demonstrated both theoretically and by emulation on a PC, meeting the required specifications. When writing this contribution (November 1998), the first generation of the FPGA implementation of the RM-TOF algorithm is realized and integrated into the complete prototype (see figure 5.1.), and is being tested in laboratory conditions. A second "tuned" generation of FPGA is scheduled for the beginning of 1999 as well as on-road tests (in collaboration with CRF and Magneti Marelli).

The final design optimization of the complete system (size, production costs, functions) will be carried out on private founding.

9. Acknowledgements

The authors wish to thank all the members of the OLMO consortium for their strong personal involvement and real collaboration spirit during the whole project, namely Mr. G.-C. Alessandretti, Mr. P.Gay, Mrs. E. Balocco and Mr. Vignani from CRF, Mr. A. Paolini, Mr. Innocenti and Mr. G. Manassero from Magneti Marelli, Mr. P. Besesty, Mr. E. Molva and Mr. Ph. Thony from LETI, Mrs. Hennig from JENOPTIK and Mrs. A.-M. Palmier and Mr. Hamidi from Renault. We also wish to thank the Office Fédéral de l'Education et de la Science (OFES), who supported the swiss contribution to the project.

100.000 Pixel 120dB Imager for Automotive Vision

T. Lulé[1], H. Keller[1], M. Wagner[1], M. Böhm[1,2], C.D. Hamann[3], L. Humm[4], U. Efron[5]
[1]Silicon Vision GmbH, D-57078 Siegen, Germany
[2]Institut für Halbleiterelektronik (IHE), Universität-GH Siegen, D-57068 Siegen, Germany
[3]Adam Opel AG, D-65423 Rüsselsheim, Germany
[4]Delphi Delco Electronics Systems, Malibu CA90265, USA
[5]Dept. of Electrical Engineering, Holon Technology Center, Holon 58102, Israel

Abstract: The newly developed locally autoadaptive image sensor LARS II (**L**okal-**A**utoadaptive**R** **S**ensor) with 368 x 256 pixels hits the demands of the automotive industry for high dynamic ranges. The splitting of the total dynamic range into a time information and an integration signal provides higher contrast images than logarithmically compressing imagers. Also locally autoadaptive sensors are inherently much less sensitive to fixed pattern noise and temperature variations. LARS employs the TFA technology where the optical detector made of amorphous silicon is deposited on top of the CMOS ASIC. The a-Si:H pin photodiode has been optimized for high sensitivity and low dark current. For the new generation of LARS imagers the pixel circuitry has been completely redesigned for lower transistor count and higher yield. It contains 17 transistors and 2 capacitors in a reduced pixel area of 40μm x 38.3μm. The chip now contains CDS and DDS capabilities for both output signals and was fabricted in the 0.8μm 2M2P CMOS process of AMS. Images taken with the first prototypes show high sensitivity, excellent local contrast and improved dark and photo fixed pattern noise. Scenes with extremely high contrasts were captured without saturation or underexposure in any part of the picture.

Keywords: Automotive Vision System, Image Sensor, TFA Technology, High Dynamic Range Image Sensor, Autoadaptive Image Sensor

1. Introduction

New automotive systems like Adaptive Cruise Control, Collision Warning and Roadway Departure Warning become more and more important in future traffic control concepts. They demand robust but cost effective camera systems with very high dynamic ranges. With conventional CCD sensors it is hard to handle higher dynamic ranges in a linear signal under the harsh environment of an automobile vehicle. They also make on-chip integration of signal processing components difficult. CMOS imagers have entered competition which overcome some of the above drawbacks. They allow systems on chip as the basis of cost effective, powerful camera but also have trouble with the dynamic range[1].

Unlike a CMOS imager the LARS II fabricated in TFA (**T**hin **F**ilm on **ASIC**) technology [2] hits all the demands of the automotive industry. The vertical integration of the device provides a fill factor of nearly 100% for both the detector and the underlying pixel circuitry. Furthermore the pixel circuitry and the optical detector are separately designed and optimized for handling strong illumination conditions in natural scenes [3]. **Fig. 1** illustrates the basic structure of the LARS II imager. The detector is formed by an a-Si:H thin film system which is sandwiched between a metal rear electrode and a tranparent front electrode. The crystalline ASIC includes the corresponding pixel circuitry underneath each detector and further the necessary peripheral circuitry for addressing and reading out the imager.

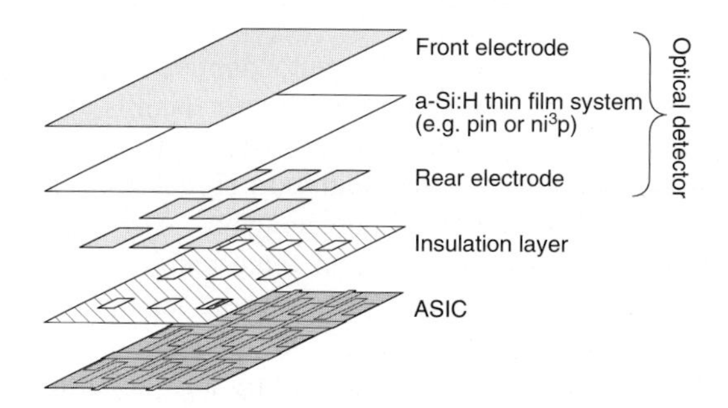

Fig. 1: Basic structure of a TFA image sensor

The following chapters investigate the demands for high dynamic range images and describe the autoadaptive concept as the solution, its technical realization and the performance measured on the LARS II imager.

2. Image Sensors with High Dynamic Range

The performance of Adaptive Cruise Control systems and other automotive vision systems depends on the quality of the images supplied by the camera system. A number of robust lane tracking algorithms have been developed in the past working well when provided with good images from the camera system. Unfortunately image sensors currently are not able to cope with real world illumination conditions. On a clear sunny day illumination levels can vary by a factor of far over 100 e.g. from outside to inside a tunnel. These 40dB combined with some 50dB of necessary contrast resolution sums up to more than 90dB of desired dynamic range [4].
CCD cameras with more than 75dB are specialized systems which work at frame rates far too low for cruise control. Also off the shelf or scientific cameras with 50dB to 75dB require highest performance components such as ADCs and are too costly and sensitive for automotive end user applications.

A common approach for the extension of the acceptable illumination range exploits the logarithmic voltage-current response of diodes or MOSFETs in sub-threshold operation [5]. These logarithmic sensors read out the compressed voltage by simple source followers which leads to very small pixels and allows random access. However, sensitivity and local contrast are poor while fixed pattern noise is temperature sensitive and amplified exponentially. This makes FPN correction inherently complicated and slow [6].

The LARS II image sensor overcomes these disadvantages by splitting the total dynamic range into two signals each with moderate dynamic range. Since both signals are stored on in-pixel capacitances self-offset-correction schemes such as CDS can be applied which effectively reduce FPN and temperature drift.

The important feature of the LARS imagers is the ability of the pixels to adapt themselves to the local illumination conditions: Every pixel optimizes its integration time to make best use of the available voltage swing without saturating. [7]

The function principle of the LARS II can best be undestood when looking at the block diagram and timing chart of **Fig. 2** and **Fig. 3**, respectively. The photo current I_{ph} is integrated to a voltage V_{signal} on the integration capacitance C_{int}. This voltage is compared to a fixed reference voltage V_{comp} at discrete points in time (rising edge of Clock). If the pixel voltage has exceeded this reference level, integration is terminated to avoid saturation. Otherwise the integration time is doubled to make better use of the voltage swing. In this way the integrated voltage is always within the range of V_{comp} to $2 \bullet V_{comp}$ for higher illumination levels.

In order to determine the chosen integration time, a voltage ramp V_{ramp} that ramps up one step after every comparison is applied to all pixels. If a pixel stops the integration it stores the actual value of V_{ramp} as a time stamp information. The value of this voltage allows the signal processor to determine which integration period the pixel had chosen. Most economically the integration times are in a powers of two series, reducing the effort for clock generation and signal processing.

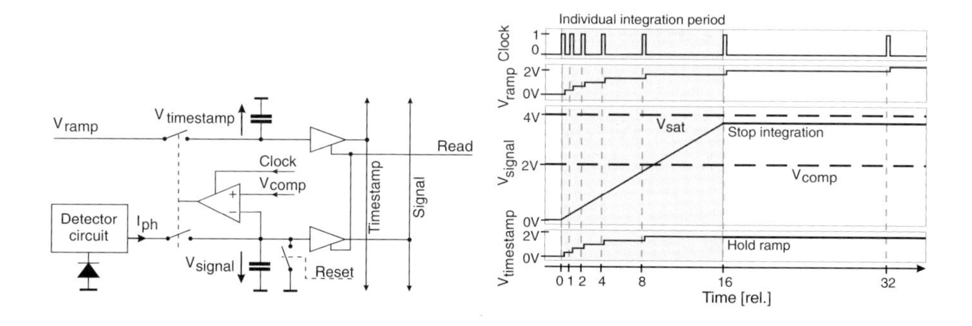

Fig. 2: Block diagram of LARS II pixel

Fig. 3: LARS II pixel timing

Both values, the time stamp and the integrated signal are read out during the following readout phase. The absolute intensity value is split into these two signals, an absolute time information (T) and intensity information (I) relative to the time information. When using powers of two timing these can be regarded as a floating point representation of the exposure (E) of the type $E = I \cdot 2^T$. Low cost components (e.g. 8bit ADCs) may be used for the processing of these signals and still dynamic ranges beyond 100dB are possible. For example 48dB linear intensity range (8bit) and a 10ms/10μs time range (60dB, encoded as 10 different time steps into a 4 bit signal) yield 108dB total dynamic range.

3. Locally Autoadaptive Sensor Design

The LARS II was designed by Silicon Vision for a double metal double poly

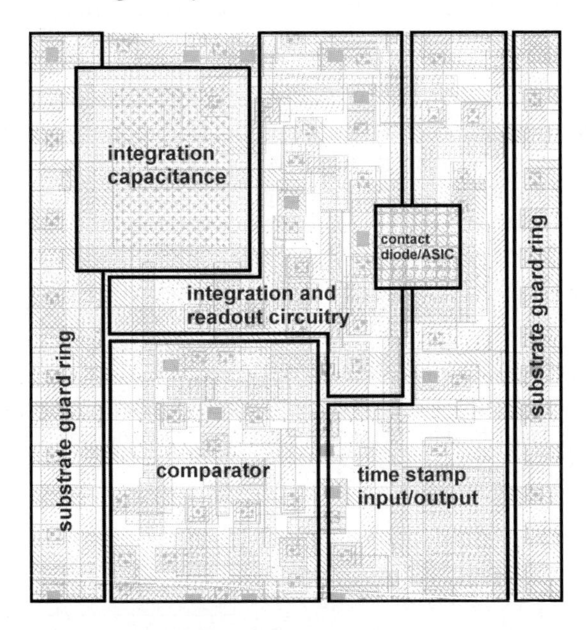

Fig. 4 LARS II pixel layout

0.8μm CMOS process for 5V. The locally autoadaptive function was realized with 17 transistors and two capacitors per pixel.

3.1 Pixel design

Fig. 4 shows the layout of one pixel circuit measuring 40μm x 38.3μm. The photocurrent of the overlaying pin-diode made of amorphous silicon is tapped by the contact in the top right quarter of the pixel and passed through clamping and multiplexing transistors onto C_{Int}. The integration capacitor realized in POLY1-POLY2 for best matching between pixels can be seen in the top left corner. The

bottom left corner contains the comparator with an inverter output stage made up of 10 transistors. The hold capacitor for the timestamp realized as a MOS varactor is shown in the right corner together with three write in and read out transistors.
Special care was taken during layout to meet the specific requirements of a mixed signal opto-ASIC for high dynamic ranges. Stray light had to be prevented from leaking into the array to insure complete shut off of integration and to suppress smear from photogenerated carriers deep in the substrate. Digital and analog circuitry was strictly separated in the top and bottom half of the pixel to reduce adverse cross coupling. Also, excessive subtrate tapping was used to supply reliable accurate backgate potentials to all transistors and to drain away photocarriers from the substrate.
The pixel fullfills the following specifications as it is designed: The integrated output voltage ranges from 4.8V at dark to 2.8V at the foldover, down to 1.8V for saturation. The pixel comparators consume a maximum of 200 nA per pixel, i.e. 19mA per chip sufficient to allow for 5μs as the shortest integration time. Power consumptions may be reduced to 1mA for applications where 100μs would be short enough. The total integration capacitance of 340fF results in a sensitivity of $0.47\mu V/e^-$ whereas the kTC noise is below 150μVrms as calculated. Since the detector delivers as much as 1pA/lux the light referred sensitivity amounts to 24mV/lux after 8ms of integration or $3V\ lux^{-1}\ sec^{-1}$

3.2 Signal Readout

The column amplifiers employ CDS which is realized by capacitive differencing (ref. **Fig. 5**). First the integrated value is read out via the column line *Column* and stored in the capacitor C_{cds} while the other capacitor plate is clamped to V_{init}. Then the pixel is reset and read out again with M2 off. The resulting potential at the gate of M3 is then Vinit plus the difference between the two readout values. In this way, ASIC fixed pattern noise and 1/f noise components are suppressed effectively. M3 and M4 represent the source follower readout to the pad drivers.

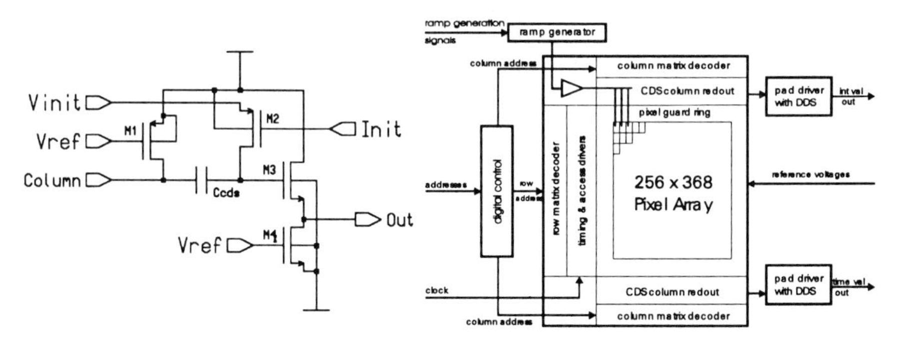

Fig. 5 Principle schematic of CDS circuit **Fig. 6** Block diagram of the LARS II imager

Equivalently the column fixed pattern noise is reduced by a DDS circuit in the pad driver. For that the column is first read out and then quickly reset to Vinit before the reset level is read out as well.
The readout circuits consume 30μA per column i.e. a total of 11mA for the column readout and 5mA per pad driver. This allows 10μs for row access and readout speeds of 14Mpix/sec with an accurate settling to 1% into 20pF load capacitance. The total power consumption sums up to 50mA at 5V.

3.3 Chip Architecture

Fig. 6 shows the architecture of the LARS II imager. The main part of the imager is the pixel matrix which is a memory like organized full custom layout. Further parts of the chip are a ramp generator, two pad drivers with DDS functionality and a comparatively small digital control unit.
For addressing the rows and columns, hierarchical matrix decoders were choosen in order to allow random access but limit the number and length of addressing lines connected to the matrix. Since the LARS II provides two output signals (time and integration value) both the column matrix decoder and the CDS column readout stage are repeated on top and bottom of the pixel array.
The ramp generator supplies the pixel matrix with the time stamp ramp. It is realized as a tapped resistive divider with 16 steps – sufficient for a time range of $1:2^{16} = 96$ dB.
Alike the pixel layout the matrix and chip design follows special „TFA Design Guidelines“ in order to ensure the high performance requirements of the imager. The pixel array for example is surrounded by a guard ring which prevents the sensitive analog nodes (diffusion regions) of the pixel circuitry from light generated carriers as well as from coupling effects through the p-doped substrate. Furthermore, a full separation of the supplies in 4 domains (Pixels, Column Sense Amplifiers, Pad drivers, Digital circuits) was implemented for

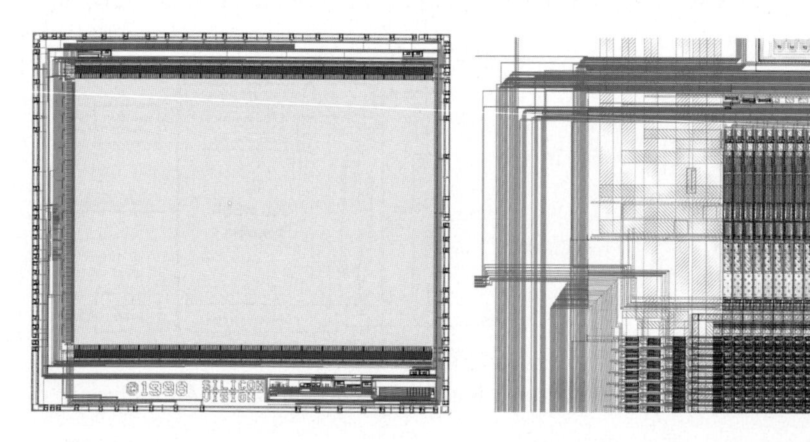

Fig. 7 Layout of the LARS II imager

Fig. 8 Detail view of the layout

suppressing cross coupling between different function blocks via the supply lines effectively. The pixel, column and digital supplies are applied to each corner of the matrix (ref. **Fig. 8**). Therefore a low impedance on the power lines is achieved which reduces voltage drops along the array. To reduce parasitic photocurrents all peripheral circuitry is shielded with the metal 3 layer, a TFA specific layer that is used as the rear electrode of the a-Si:H photodiodes (ref. **Fig. 1**).
The layout of the LARS II chip wich has a total size of 16.5 mm x 14.9 mm is shown in **Fig. 7**. It includes the main matrix (14.6 mm x 11.9 mm, light grey area plus repeating structures at the top, bottom and left side), a ramp generator (top left), a digital control unit (left), two DDS pad drivers (top right, bottom right), a large block containing different test circuits (bottom) and a pad ring. **Fig. 8** depicts the upper left corner of the pixel matrix in detail. One can see the pixels surrounded by a guard ring and the neighboring cells for row and column access. Also the six different voltage supplies can be seen clearly.
Fig. 9 and **Fig. 10** show the die in a 100 pin PGA package and the camera board, respectively. The board contains all support circuitry including timing generation, supply and reference regulators. The 12 bit AD converters running at 20MHz deliver full frames at 60Hz directly to a signal processor. Also a PAL/NTSC compliant BAS signal is generated for direct monitoring on a TV. Most of these functions will be included on chip in succeeding versions of the sensor for the camera to be more compact and cost effective.

Fig. 9 LARSII imager in PGA package

Fig. 10 Camera board for LARSII with ADCs

4. Image Sensor Performance

Measurements were carried out on the LARS II with white light at 50Hz frame rate and 3ms integration period. Images are only dark level corrected.
With 1.8V lx^{-1} sec^{-1} the output referred sensitivity is lower than calculated for the pixel due to source follower gains smaller than unity. The noise floor is below 2mV_{rms} whereas the full scale signal is 2V which is equivalent to 60dB of dynamic range. On the other hand the integration times may vary between 10ms and 10μs which increases the illumination range by another 60dB, resulting in 120dB of total dynamic range.

Fig. 11 through **Fig. 15** demonstrate the performance of the LARS II autoadaptive imager. In **Fig. 11** the integration time was fixed to 3 ms whereas **Fig. 12** shows the imager with fixed integration time of 200μs. The contrast in the scene is so high that no integration is adequate to prevent overexposure or loss of detail in the dark regions.

Fig. 11 Image scene with 3ms integration time (overexposure)

Fig. 12 Image scene with 200μs integration time (underexposure)

An imager with pixels that can adapt to its local intensity is able to cope with such a contrast. **Fig. 13** and **Fig. 14** show the integration image and time stamp image taken with the LARS II switched to autoadaptive operation. For lower intensities the integrated image looks like a normal grayscale image. However, for increasing intensities the integrated value switches back to darker values multiple times. The time image clearly shows regions of constant but reduced integration times (brighter is shorter) whose edges match the bright-dark transition edges of the time image.

Fig. 13 Integration image taken with LARSII

Fig. 14 Time stamp image taken with LARSII

Both images **Fig. 13** and **Fig. 14** where composed together to yield the reconstructed scene of **Fig. 15**. The detailed information in the dim and bright areas are clearly visible. The contrast had to be compressed by gamma correction to be able to see the full information in a paper print. The sharp image of the bulb filament shows the blooming resistance of the pixels.

Fig. 15 Effective image, reconstructed from time and integration information

5. Conclusions

Automotive vision systems require image sensors with very high dynamic range (>100dB) and blooming resistance at low cost. State of the art CCD and CMOS sensors with high dynamic range (still less than 90dB) are very expensive and stress sensitive, whereas logarithmic sensors suffer from excessive, temperature dependent fixed pattern noise. The LARS image sensor family perfectly meets the demands of the automotive industry for high dynamic range in a robust, cost effective image system. Their autoadaptive pixels adapt to the locally impinging illumination intensity by varying their integration time in realtime. In this way the integrated voltage makes best use of the voltage swing without saturating. The total dynamic range is thereby split into two signals with moderate dynamic range (<60dB) which are easy to process with low cost analog components and signal processors.

A locally autoadaptive image sensor LARS II with 368 x 256 pixels was fabricated. The sensor contains CDS and DDS readout circuits as well as random access address decoders and a time stamp ramp generator. Images taken from scenes with high contrast show the power of the locally autoadaptivity to cope with high dynamic ranges. This makes the LARS II an ideal solution for automotive vision systems that have to deal with high contrast natural scenes.

References

[1] C. Cavadore, J. Solhusvik, P. Magnan, A. Gautrand, Y. Degerli, F. Lavernhe, J. Farré, O. Saint-Pé, R. Davancens, M. Tulet, "Design and characterization of CMOS APS imagers with two different technologies", Solid State Sensor Arrays: Development and Applications II, 26.-27. Jan. 1998, p. 140, San Jose, California, ed. M. Blouke, SPIE, Vol. 3301, 1998

[2] B. Schneider, P. Rieve, M. Böhm, "Image Sensors in TFA (Thin Film on ASIC) Technology", Handbook of Computer Vision and Applications, Academic Press, Boston, 1998

[3] T. Lulé, H. Fischer, S. Benthien, H. Keller, M. Sommer, J. Schulte, P. Rieve, M. Böhm, "Image Sensor with Per-Pixel Programmable Sensitivity in TFA Technology", H. Reichl, A. Heuberger, Micro System Technologies '96, VDE-Verlag, Berlin, p. 675, 1996

[4] M. Böhm, F. Blecher, A. Eckhardt, B. Schneider, S. Benthien, H. Keller, T. Lulé, P. Rieve, M. Sommer, R.C. Lind, L. Humm, M. Daniels, N. Wu, H. Yen, U. Efron, "High Dynamic Range Image Sensors in Thin Film on ASIC Technology for Automotive Applications", Advanced Microsystems for Automotive Applications (AMAA), Berlin, 26./27. 3. 1998

[5] M. Loose, K. Meier, J. Schemmel, "CMOS image sensor with logarithmic response and self calibrating fixed pattern noise correction", EUROPTO Conference on Advanced Focal Plane Arrays and Electronic Cameras II, Zürich, Switzerland, May 1998, SPIE, Vol. 3410, p. 117, 1998

[6] S. Fischer, N. Schibli, F. Moscheni, "Design and Development of the Smart Machine Vision Sensor (SMVS)", EUROPTO Conference on Advanced Focal Plane Arrays and Electronic Cameras II, Zürich, Switzerland, May 1998, SPIE, Vol. 3410, p. 186, 1998
[7] "Bildsensor mit hoher Dynamik", patent pending, PCT/EP 97/95 306, 29.9.1997

Development of RF Technology for Automotive Radar and Mobile Broadband Communication

Frank Rehme, Holger Meinel, Wilhelm Hager
DaimlerChrysler Aerospace AG
Woerthstrasse 85; D-89077 Ulm
Pone: ++49/731/392-4540; Fax: ++49/731/392-5465
frank.rehme@vs.dasa.de

Keywords: Automotive Radar, Broadband Communication, Antennas, Packaging

After digital electronics entered the automotive industry years ago it is now the radio frequency (RF-) electronics with frequencies far above 2GHz which is starting to be applied in cars and other means of transportation. Since 1975 many companies, institutions and laboratories experimented with various millimeter-wave radar systems for automotive applications. The aim was to develop Obstacle Warning Radars (OWR), warning the driver of potential hazards in his path, and Autonomous Intelligent Cruise Control (AICC). Further approaches are radar systems for measuring the velocity or ice warning radars.
Up to now only the Intelligent Cruise Control is commercially available now since the real-time analysis of the received data is highly complicated and needs to be extremely reliable, especially in case of an Obstacle Warning System which must include a suitable man-machine-interface. This leads to a cooperative driving and is therefore the ultimate, but long-term future.
In the field of broadband communication the development of terrestrial Point to Multipoint Systems anticipate the development of future *mobile* broadband communication systems. Nowadays transceiver front ends at about 30 or 40 GHz are reaching a price level, which allows the development of products for the consumer market capable of transmitting data at a rate of several hundred Mbit per second. In the future these systems will be used for „Video on Demand", broadband Internet access, videoconferencing and the like. They consist of a flat antenna being mounted on the roof top looking at a base station much smaller than a base station of common GSM mobile phones.
The next step in in the field of wireless broadband communication is the transmission of data from a personal computer via a small transceiver to satellites, cruising airplanes or unmanned zeppelins. These flaying or orbiting base stations distribute the data to several hubs of Internet providers telephone companies or Videodistributors.
Broadband Communication via satellite is also possible from and to vehicles or other means of transportation. It is most easily accomplished for airplanes („Internet in the skies"), which always have a free „line of sight" to the satellite. In some years broadband communication will be offered to the passenger of commercial airplanes. Trains and vehicles will follow this trend. Because the transmission of

data at these high frequencies dose not tolerate obstacles between antenna and satellite first broadband systems for vehicles will be workable (like Astra links for caravans) only when the car has stopped. Therefore first possible applications are life transmission of video data or for instance mobile "video on demand".
For all mentioned products three factors are critical: the GaAs-chips, generating the RF-power, their packaging and the realization of filters and antennas. While a drastic cost reduction of the GaAs-chips has taken place in the last years mainly due to a higher integration of the circuits the development of cheep packaging methods for GaAs-components is only beginning.
In recent years some work was done at DaimlerChrysler-Aerospace AG on forming Packages for RF-modules as well as antennas and waveguide filters by application of common casting methods for metals, ceramics and plastics. Since plastic injection molding is most widely used to realize highly complicated 3 dimensional geometries injection molding was predominantly tested. The second method examined was powder injection molding (PIM), which allows to form complex parts from metals as well as ceramics. PIM is especially of interest for heatsinks incorporating DC and RF feedthroughs. Preferred materials for this application are Fe, CuW, CuMo, Al/SiC or ceramic materials like AlN.

Tests with injection molded parts show, that sufficient small tolerances in the mechanical dimensions can be achieved which are necessary to build waveguides at about 60 GHz from metalized plastic parts. Figure 1 shows a test structure for an automotive radar.
The change from milled aluminum to metalized plastics of corse is followed by significant changes in many other technologies needed to complete the RF-front end: For example the way the parts are bonded has to regard the special properties and limits of plastics. Therefore gluing and soldering techniques have been examined to bond metalized plastics with plastics as well as plastic parts with metals. Soldered waveguide filters from liquid crystal polymers operating at 28 GHz have shown losses similar to machined filters from aluminium. Also antennas build from a metalized plastic part and a slotted metal sheet have been successfully bonded by soldering (figure 2). The injection molded part from liquid crystal polymer together with the metal sheet are forming a slotted waveguide antenna operating at 28 GHz. The antenna is part of a broadband transceiver front end demonstrator being currently under development.
Besides bonding and assembly the RF-design has to be altered in order to utilize plastics in RF-applications. For that reason new design guidelines have to be developed. Subsequently the whole system design has to be revised and adopted to the molding technology. The advantage of utilizing injection molding lies in decreasing the cost per part and in accomplishing a higher level of integration of functions and therefore a reduction of the number of parts necessary. Additionally the overall wait can be reduced drastically. Figure 3 shows a test structure build for a flat automotive radar sensor. The antenna and the RF-part is integrated in the housing made of plastics.

Acknowledgement

The research was supported by the German Federal Ministry for Education, Science, Research and Technology under project number 16 SV 416 / 9.

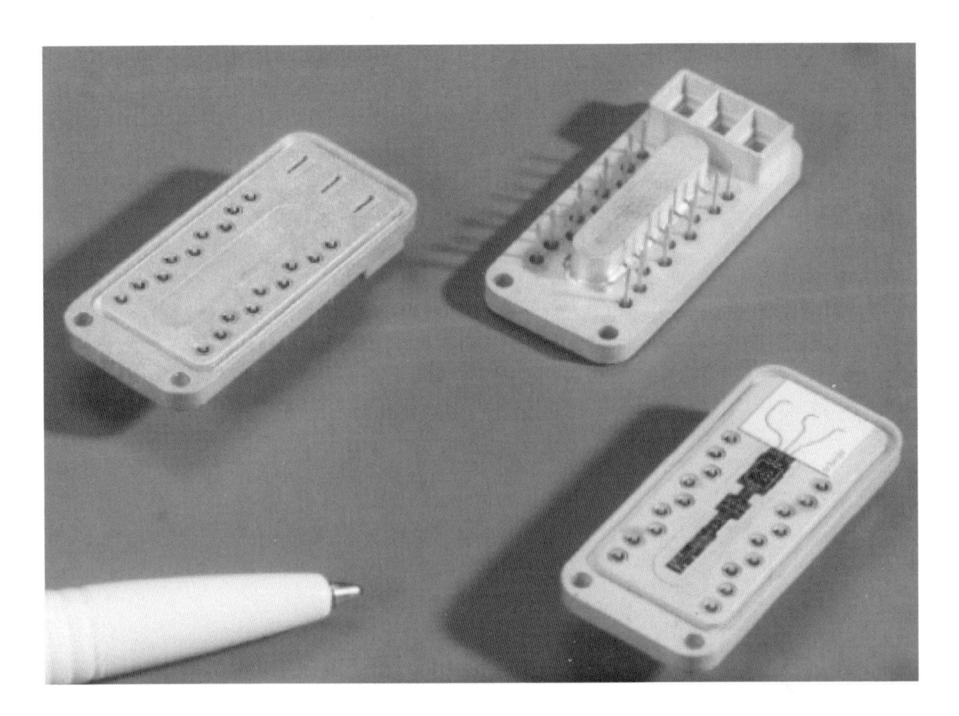

Fig. 1: Injection molded test structure for automotive radar with waveguides. The metalized plastic parts are made from polyethersulfone.

Fig. 2: Waveguide antenna from metalized liquid crystal polymer

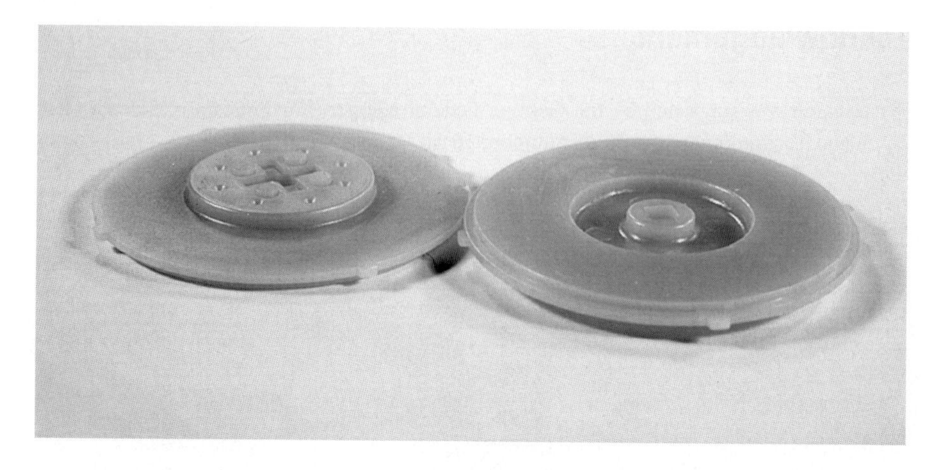

Fig. 3: Injection molded test structure for a radar sensor made from polyethersulfone

A Low cost micro-inertial and flow sensors based on the Direct Integration technology

Dan Haronian, Department of Interdisciplinary Studies, Faculty of Engineering
Tel-Aviv University, Ramat-Aviv 69978, Tel-Aviv, Israel
Tel:+972-3-640-6414, Fax:+972-3-641-0189, Email:haronian@eng.tau.ac.il

Abstract: The penetration of a MEMS technology into the automobile industry strongly depends on its cost. Such penetration can be accomplished by using a low cost MEMS technology that can be easily integrated with conventional silicon based microelectronics. The Direct Integration technology, presented in this paper, offers a low cost integration of displacement sensing of micro-beams with conventional microelectronics. The micro-beams, with typical cross section of 2 mm x 20 mm, are fabricated from single crystal silicon using SCREAM process, and have an in-plane degree of freedom. The sensors, made of *pn* diodes and nMOS transistors, are fabricated at the beam's support close to their root. When the beams deform stresses develop at their root that modulate the band gap energy of the silicon. This modulation affects secondary properties of the sensors at the beam's root, that is translated into electrical signals, that are further processed by on chip ICs. The sensors and the ICs were fabricated by a conventional microfabrication facility and the mechanical elements were integrated subsequently. This integration adds only one lithography step to the fabrication steps of the IC insuring low cost integration.

Keywords: Inertial Sensors, SCREAM Process, Flow Sensors

1. Introduction: The planarity nature of microfabrication technology and its affects on MEMS technologies

Planarity is a fundamental nature of microfabrication technology. This property is responsible to the fact that different element in VLSI such as diodes, transistors, capacitors etc., are placed and interact in the plane of the wafer. This planarity property is inherited by the MEMS technology, and many MEMS are comprised of element such as sensors and actuators that are fabricated and interact in the plane of the wafer and therefore have an in-plane Degree Of Freedom (DOF) (see Fig. 1a). For example, the micro x-y-z stage carrying an STM tip[1], the vibrating gyroscope [2] or the micro-gear [3] are all MEMS with several mechanical components interacting with each other in the plane of the wafer.

In-spite of this planarity nature, MEMS with out-of-plane DOF can also be found (see Fig. 1b). The micromachined microphone made of a suspended membrane over a sealed cavity [4], and the pendulum accelerometer fabricated using wet etch of silicon [5], are examples of devices with out-of-plane DOF. Still because of the planarity nature of the microfabrication technology these MEMS have low mechanical integration abilities and therefore are capable of performing only simple tasks.

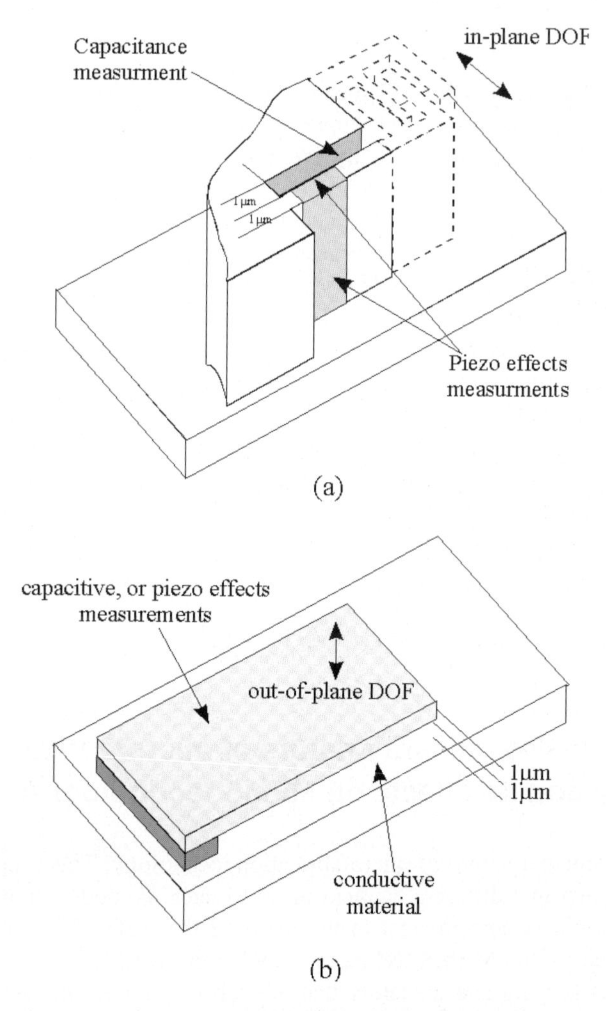

Fig 1: A schematic description of a beam with in-plane (a) and out of plane (b) degree of freedom

The disadvantage of the planarity nature of the microfabrication technology becomes an advantage when it comes to integrating a displacement sensor for elements with out-of plane DOF. Since these elements move out-of the plane it is relatively easy to fabricate sensors such as capacitive sensors, piezoresistive, or piezoelectric sensors on these planar elements. The movement of the micromachined microphone for example, can be sensed capacitively by coating the membrane and a counter close plane with metal, or it can be sensed by coating piezoresistive or piezoelectric materials on the membrane. On the other hand, the advantage of the planarity nature of the microfabrication technology for elements with in-plane DOF becomes a disadvantage when it comes to integration of a displacement sensor. Since elements with planar DOF move in the plane of the wafer the traditional sensing concepts should be fabricate on their sidewalls. Not only this task is not trivial with conventional microfabrication technology, typically the overall area of the sidewall is too small to be effective. This is why comb structures, that increase the overall sidewall area, are used for capacitive sensing (see fig 1a).

We can conclude that: *MEMS with out-of plane DOF have low mechanical integration capabilities and high sensing integration capabilities, while MEMS with in-plane DOF have high mechanical integration capabilities and low sensing integration capabilities*.

The disadvantages of the two technologies increase their cost and complexity. Therefore, in order for these technologies to become attractive these disadvantages should be eliminated. To increase the mechanical integration ability of a MEMS technology with out-of-plane DOF one need to fabricate suspended elements one on top of the other either using multiple chip technology or using non-planar microfabrication technology. Multiple chip technology is relatively expensive while non-planar microfabrication technology is not available yet. On the other hand to increase the sensing integration ability of a MEMS technology with in-plane DOF one need to associate the in-plane mechanical movement with some sensible physical property *without adding fabrication complexity*. The Direct Integration (DI) technology does exactly that: It was found, using a finite element analysis, that the stress developing in a fixed-free single crystal beam during deformation extends into the support of the beam The DI uses this stress to modulate the electrical properties of solid state sensors such as *pn* diodes, MOS or bipolar transistors integrated at the root of the beam. This paper describes the potential of this technology by demonstrating the performances of accelerometers and flow sensors with in-plane DOF that are integrated with *pn* diodes and MOS transistors.

2. The effect of stress on the electrical properties of semiconductors

Stress affects the electrical properties of semiconductors by modulating the band gap energy. This affect was observed in the early 1960's by studying the performance of *pn* diodes and bipolar transistors [6, 7] under mechanical stress. The stress coefficient of the band gap energy was found to be in the order of 10^{-12} eVcm^2/dynes [8]. This band gap energy modulation affects secondary properties of solid state devices. These secondary properties are the intrinsic concentration, the charge carrier concentration, the built-in *pn* junction potential, the junction width, and the mobility of holes and electrons through the piezoresistance tensor. A detailed discussion of the affects of stress on the I-V relation of pn diodes and MOS transistors can be found in Ref. [9]. Table 1 summarizes the effects of stress on the properties of silicon.

Table 1: The effect of stress on the properties of silicon.

Property	stress function
Intrinsic carrier concentration, n_i	$n_i^2(\sigma) = n_{i0}^2 e^{\frac{E_g+10^{-12}\sigma}{kT}}$
Density of carriers	$n_1(\sigma) = n_{10} e^{\frac{E_g 10^{-12}\sigma}{2kT}}, p_1(\sigma) = n_{10} e^{\frac{E_g+10^{-12}\sigma}{2kT}}$
built-in *pn* junction potential	$\Phi_b(\sigma) = \frac{kT}{q}\left(Ln\left(\frac{N_a N_d}{n_{i0}^2}\right) - \frac{E_g+10^{-12}\sigma}{kT}\right)$
pn junction width	$W_j(\sigma) = \sqrt{\frac{2\varepsilon_s\varepsilon_0(N_d+N_a)}{qN_dN_a}(\Phi_b(\sigma)+V)}$
Mobility	$\mu(\sigma) = \mu(1+\sigma\pi_t)$ (For transverse stress)

3. The scream process

The SCREAM [10] (Single Crystal Reactive Ion Etching and Metalization) micro-fabrication process is capable of fabricating deep structures with very high aspect ration (about 1:150). Still this process suffers from one major disadvantage. Before describing this disadvantage the SCREAM process will be described. This process (see Fig. 2) starts with a deposition of SiO_2 (1 mm) layer The device image is transferred to the wafer top using optical lithography, and SiO_2 are etched using RIE with CHF_3 (steps B,C). The image is further transferred into the silicon bulk using an ICP etch using Bosch process (step D) The depth of is etching step is typically in the range of 10-100 mm. Next in step E the pattern is passivated using PECVD of SiO_2, and the floor oxide is etched using RIE of CHF_3 (step F). In step G the pattern is etched again by ICP machine using the Bosch process. This step exposes bare silicon under the passivated silicon, that is etched in step H by an isotropic RIE with SF_6. This results with free structures suspended 3-4 mm from the substrate. As can be seen there is an under-cut in unreleased regions. Therefore if the structure is metelized using sputter machine the metal coating does not reaches the under-cut regions and therefore the suspended structures are electrically isolate. The simple structure that is demonstrated in this figure contains a suspended beam (fixed-free: fixed from one side and free from the other side) placed close to a fixed wall (stationary pole). Therefore by applying a voltage between the beam and the fixed wall electrostatic forces are generated that attracts the free beam towards the fixed wall. As can be seen this process is based on one lithography step, and therefore can lead to a very low cost devices.

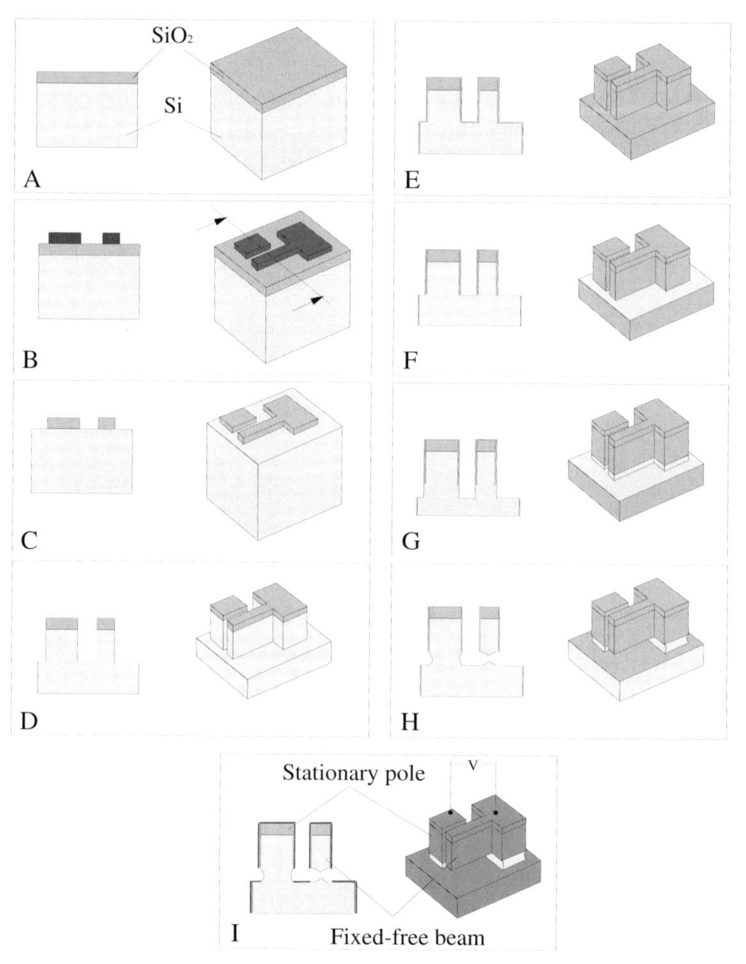

Fig. 2: The SCREAM process.

4. Devices configuration

Several devices were integrated with *pn* diodes and MOS transistors. These devices were fabricated using SCREAM process except for the deep chlorine etch. This etch was replaced by an Inductive Coupling Plasma (ICP) etch using the Bosch process, that allows fabrication of structures with larger aspect ratio Figure 3a is an SEM picture showing a single beam integrated with two diodes fabricated symmetrically in the support of the beam. The two p^{++} notations represent the conductors that are connected to the two p^{++} regions of the diodes while the single 'n' notation represents the conductor that is connected to the common 'n' region of the diodes. The common 'n' region is grounded while the p^{++} regions are connected to a current source and to a differential amplifier. When the beam deforms to one side a compressive stress develops in this side while a tensile stress develops in the other side. Therefore opposite stress effects are expected at the two diodes. These effects change the voltage drop applied on the diodes, and their difference is amplified by the differential amplifier. Figure 3b is an SEM picture showing a single beam integrated with an nMOS transistor.

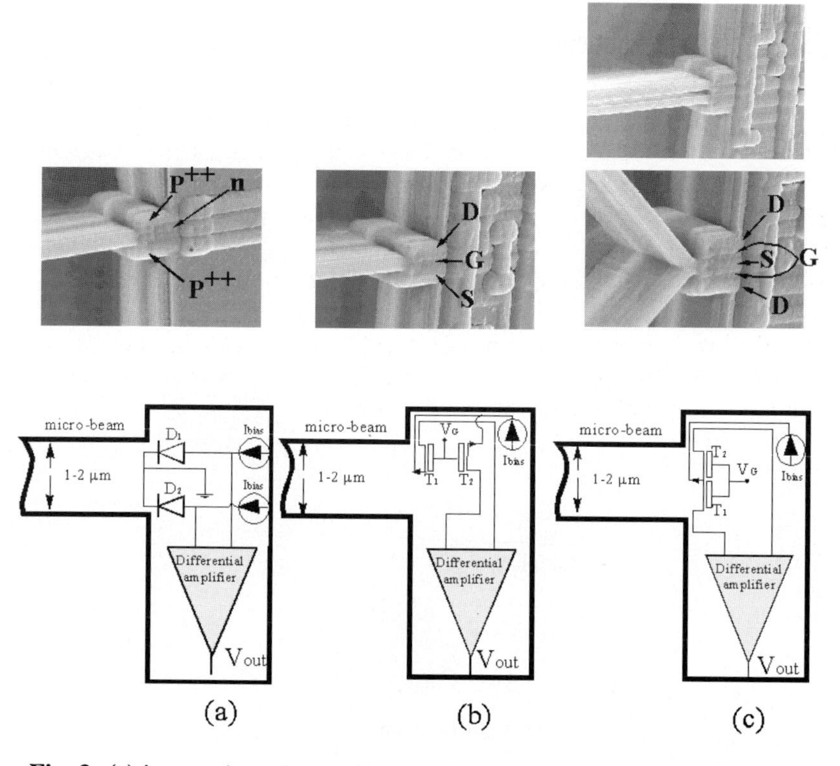

Fig. 3: (a) integration of two diodes with single crystal silicon beam, (b) integration of single transistor with single crystal silicon beam, (c) integration to two single crystal silicon beams with two transistors

The S, D, G notations represent the source drain and gate of the transistor. The sources of this transistor and of an identical transistor, located outside the stressed area, are connected in parallel to a current source, their drains are fed into a differential amplifier, and their gates are held at the same potential. In this configuration the difference in the current flowing through the channels of the two transistors is amplified by the differential amplifier. Note that the stressed transistor is located off-axis. This is done in order to avoid opposite electrical effects as a result of tensile and compressive stresses that are developing on the two sides of the beam. In figure 3c two beams with common support are integrated with two nMOS transistors. In the upper picture two parallel beams are connected to the transistors such that each beam is connected to the channel of each transistor, and in the lower picture two vertical beams are connected to the transistors such that the channels of the two transistors are located at the edges of the beam. As in the previous case the sources of the two transistors are connected in parallel to a current source, their drains are fed into a differential amplifier, and their gates are held at the same potential.

5. Devices testing

Using these sensors several devices were fabricated and tested using flow and acceleration, and by using acoustical and mechanical shocks. Figure 4 is a picture showing the experimental setup. The chip containing the devices and the supporting IC is mounted on a small board that is connected electrically to a larger board and mechanically, through a supporting beam, to a shaker. The large board contains additional amplifiers and electronics to select the output of one device out of the several fabricated on the chip. A calibration accelerometer is also connected to the supporting beam and is used to calibrate the response of the tested device. For the flow experiment a small tapered pip with 200 mm output diameter was used to flow nitrogen over the device.

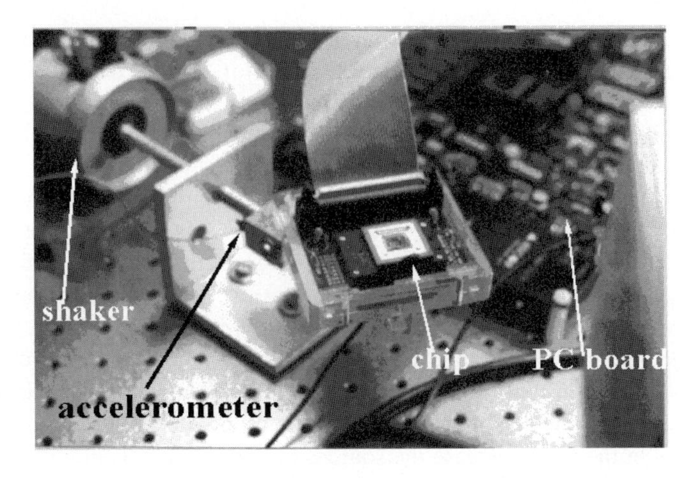

Fig. 4: The experimental setup

Figure 5 shows a proof mass supported by four beams. One of the beams is integrated with nMOS transistor as shown in Fig. 3b. The proof mass is about 1.7 x 10^{-8} kg and the overall beams spring constant is about 4.6 N/m. The calculated resonance frequency of this lumped system is found to be 2.6 kHz.

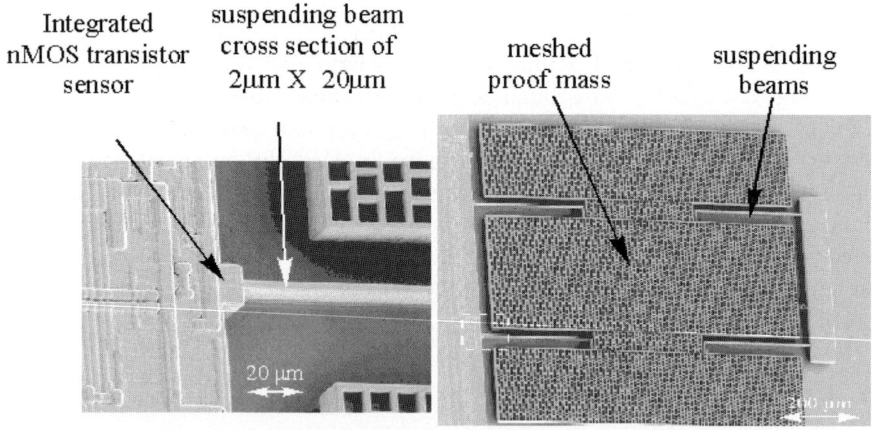

Fig. 5: A proof mass supported by four beams where one beam is integrated at its root with a single nMOS transistor as shown in Fig. 3b

Figure 6 shows the response of the sensor to acceleration of 0.03 g at different frequencies revealing its resonance characteristic.

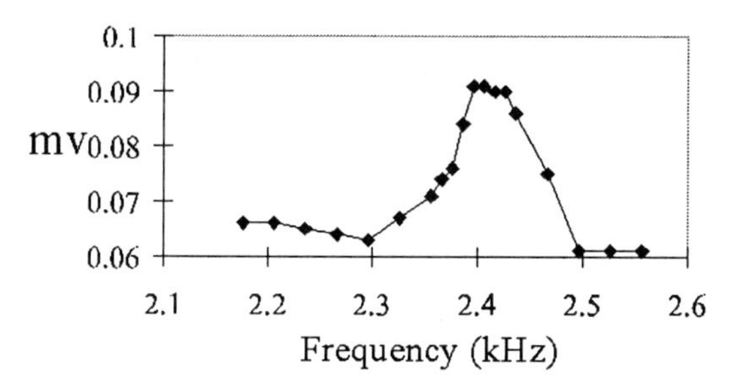

Fig. 6: The resonance response of the sensor in Fig. 5

Figure 7 shows the response of the sensor to acceleration at different amplitudes at 1 kHz. The calculated off resonance sensitivity is found to be 0.8 mv/g.

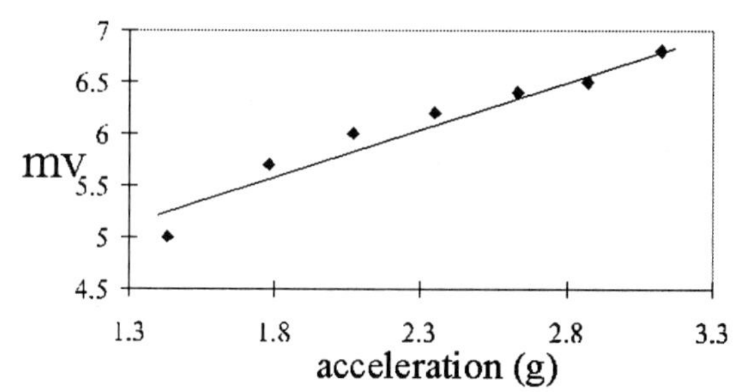

Fig. 7: The response of the sensor in Fig. 6 to acceleration at different amplitudes

Figure 8 shows an L shaped sensor with proof mass at its end, integrated with two diodes as shown in Fig. 3a. The proof mass is 2.7 x 10^{-10} kg and the first bending mode of the beam has a spring constant of 0.2 N/m. The calculated resonance is found to be 4.4 kHz.

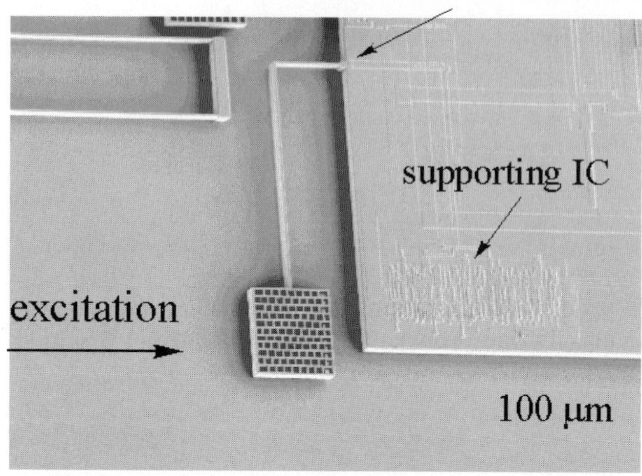

Fig. 8: An L shaped beam with proof mass at its end integrated with two diodes as shown in Fig. 3a.

Figure 9 shows the response of the sensor to acceleration of 0.03 g at different frequencies revealing it resonance characteristic.

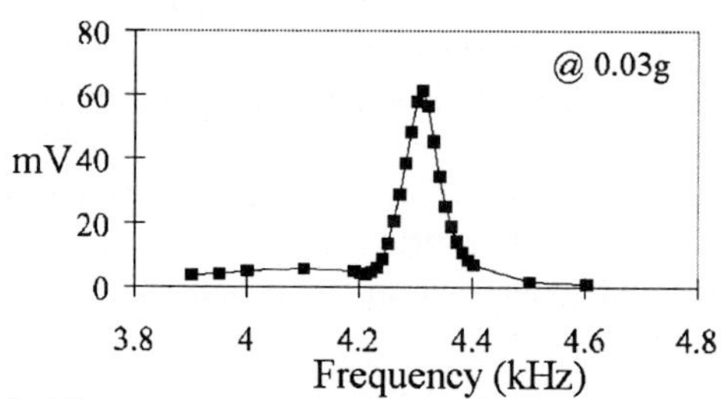

Fig. 9: The resonance response of the sensor in Fig. 8.

Figure 10 shows the response of the sensor to acceleration at different amplitudes at 1 kHz and 4 kHz. The calculated off resonance sensitivity is found to be 326 mv/g.

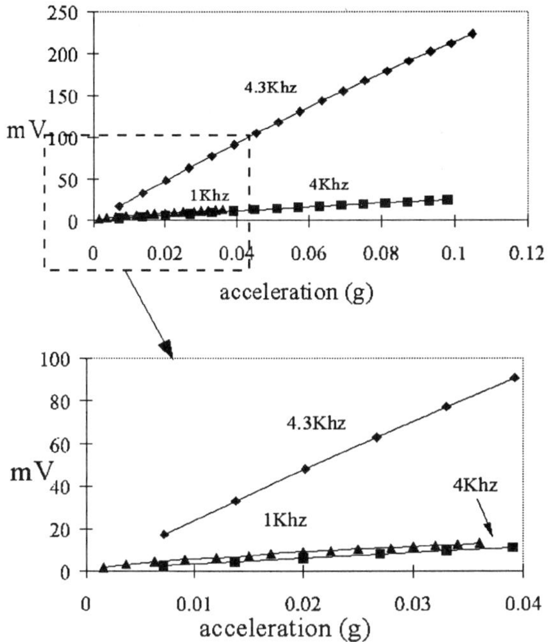

Fig. 10: The response of the sensor in Fig. 8 to acceleration at different amplitude.

Figure 11 shows the response of the sensor to acoustical and mechanical shocks. A similar resonance characteristic is excited by these shocks, that is slightly shifted to lower frequencies.

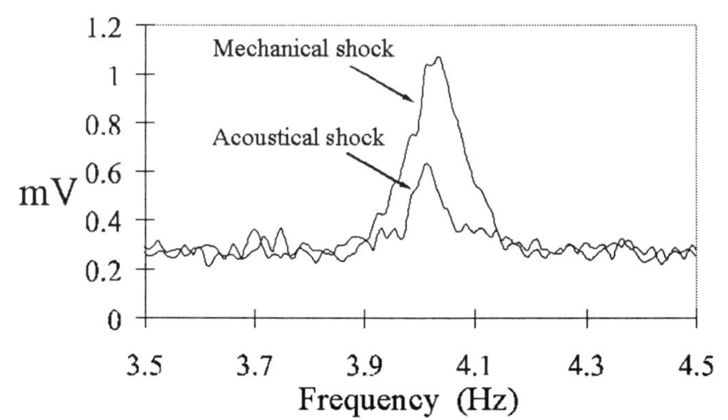

Fig. 11: The response of the sensor in Fig. 8 to mechanical and acoustical shocks.

A little more complicated configuration is shown in Fig. 12. Here a cantilevered coiled beam with proof mass at its end is integrated with a single nMOS as shown in Fig. 3b and with two *pn* diodes as shown in Fig. 3a.

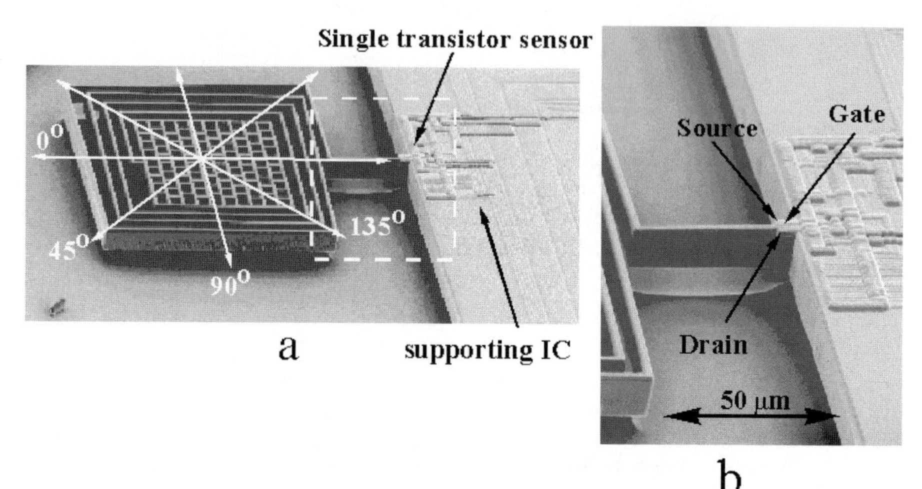

Fig. 12: A cantilevered coiled beam with proof mass at its end, integrated with two pn diodes or with single transistor as shown in Fig. 3

Figure 13 shows the response of the sensor integrated with the two *pn* diodes to acceleration of 4 g from four directions (see Fig. 12). These results show that while the sensors described in figs. 5 and 8 are sensitive to *one direction* this sensor is sensitive to *in- plane* excitation.

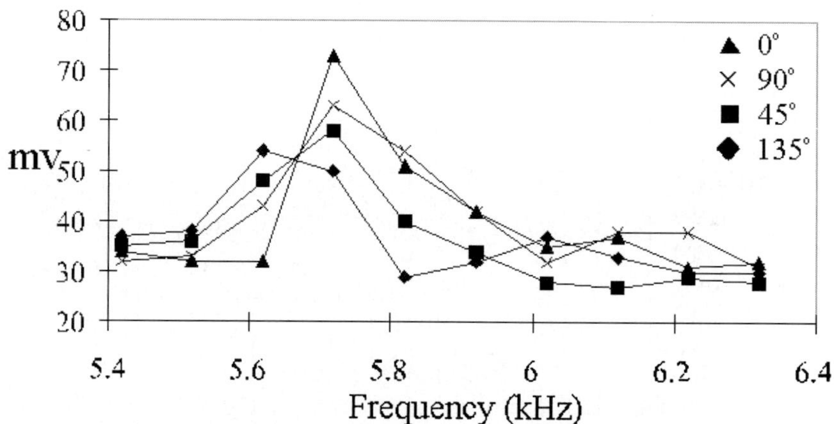

Fig. 13: The response of the sensor in Fig 12, integrated with single nMOS transistor, to acceleration in different directions

The L-shaped sensor and the coiled sensor were also excited with nitrogen flow. This flow forces the sensor to resonate at its natural frequency, and therefore similar to the floating element flow sensor [11, 12], this sensing concept is based on the interaction of the flow with the mechanical properties of the sensors. Nevertheless, unlike the floating element sensor, this sensing concept is based on resonance response rather than on deformation. Potentially, resonance response may have higher resolution as the noise floor is typically lower. Figure 14 is the response of the L shaped sensor to flow at different flow rates. As this figure shows, both the amplitude and the resonance frequency are effected by the flow rate.

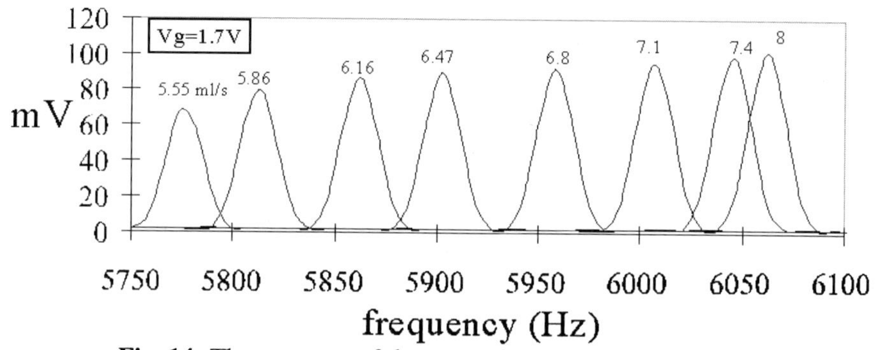

Fig. 14: The response of the sensor in Fig. 12, integrated with single nMOS transistor, to flow at different flow rates

Figures 15a and 15b show the amplitude and resonance response of the sensor to different flow rates. The calculated amplitude and frequency sensitivities are found to be 58 mv/ml/s and 250 Hz/ml/s respectively.

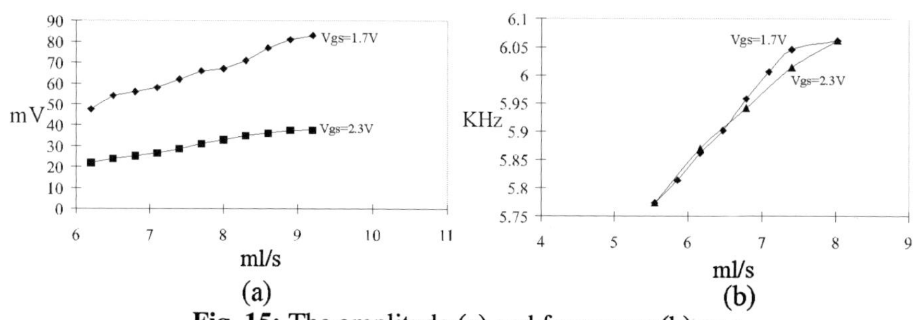

Fig. 15: The amplitude (a) and frequency (b) as a functionof flow rate of the sensor in Fig. 12

Although the devices described above are simple they demonstrate the abilities of the DI technology. This technology can be easily integrated into micro-actuators. Figure 16 shows the integration of a micro-stage with an nMOS transistor,

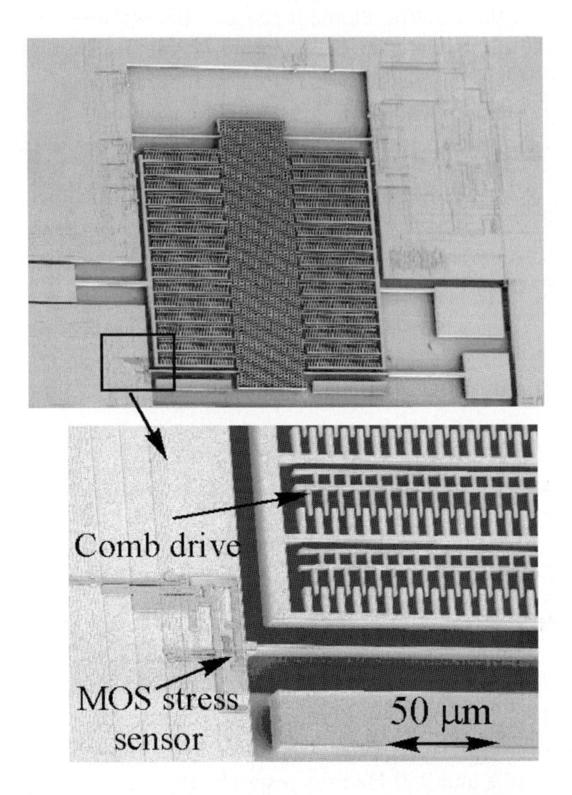

Fig. 16: A micro-stage integrated with single nMOS transistor

and Fig. 17 shows the integration of vibration gyroscope with an nMOS transistor. These and other configurations are currently being tested.

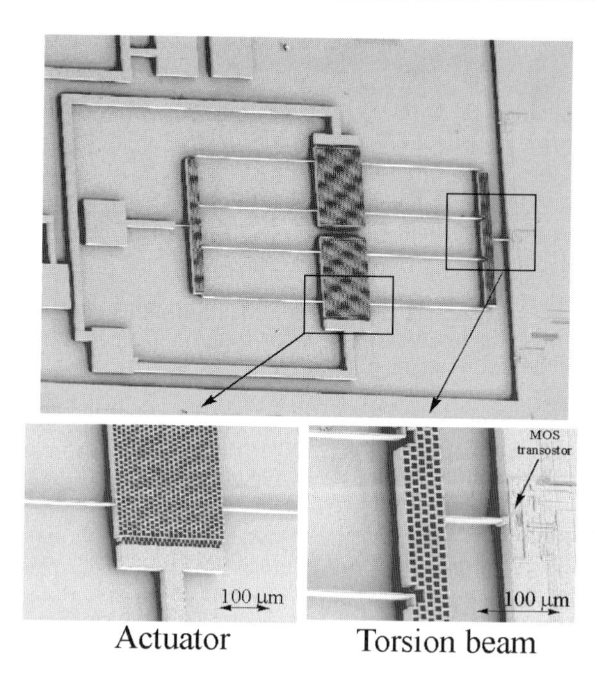

Fig. 17: A vibrating rate gyroscope integrated with single nMOS transistor

6. Conclusion

The direct integration technology was demonstrated by integration with in-plane DOF devices. Accelerometers with sensitivity as high as 326 mv/g and flow sensor with amplitude and frequency sensitivities as high as 58 mv/ml/s and 250 Hz/ml/s respectively were demonstrate. The mechanical parts of the sensors were fabricated using SCREAM process and therefore this fabrication step adds only one lithography step to the overall fabrication process. This integration is therefore very efficient and cheap. Although the DI concept is demonstrated using the SCREAM process, it may also be applied to other microfabrication technology such as those that are based on SOI.

References

1. Xu Y, MacDonald N C, Miller S A (1995) Integrated micro-scanning tunneling microscope. Applied Physics Letters 67: 2305-2307.

2. Maenaka K, Fujita T, Konishi Y, Maeda M (1996) Analysis of a highly sensitive silicon gyroscope with cantilever beam as a vibrating mass. Sensors and Actuators A 54: 568-573.

3. Legtenberg R, Berenschot E, Elwenspoek M, Fluitman J H (1997) A fabrication process for electrostatic microactuator with integrated gear linkages. Journal of Microelectromechanical Systems 6:234-241.

4. Yazdi N, Najafi K (1997) An all-silicon single-wafer fabrication technology for precision microaccelerometer. In: Tech. Dig. 9th Int. Conf. Solid-state sensors and actuators (Transducers '97) Chicago IL pp 1181-1184.

5. Bergqvist J, Rudolf F (1994) A silicon condenser microphone using bond and etch-back technology. Sensors and Actuators A 45:115-124.

6. Wortman J J, Hauser J R, Burger R M (1964) Effect of Mechanical Stress on p-n Junction Device Characteristics. J Appl Phys 35:2122-2131.

7. Wortman J J, Hauser J R (1966) Effect of Mechanical Stress on p-n Junction Device Characteristics, II. Generation-Recombinant Current. J Appl Phys 37:3527-3530.

8. Goetzberger A, Finch R H (1964) Lowering the breakdown voltage of p-n junction by stress. J Appl Phys 35:1851-1854.

9. Haronian D (1998) Direct Integration of solid state stress sensors with Single Crystal Micro-Electro-Mechanical Systems: Theory and applications. Submitted to the Journal of Microelectromechanical Systems.

10. Shaw K A, Zhang Z L, MacDonald N C (1994) SCREAM 1: A Single Mask, Single-Crystal Silicon, reactive ion etching process for microelectromechanical structure, Sensors and Actuators A, 40: 63-70.

11. Roche D, Richard C, Eyrud L, Audoly C (1996) Piezoelectric bimorph bending sensor for shear-stress measurement in fluid flow. Sensors and Actuators A 55:157-162.

12. Su Y, Evans A G R, Brunnschweiler A (1996) Micromachined silicon cantilever paddles with piezoresistive readout for flow sensing. J Micromech Microeng 6:69-72.

Novel Rotation Speed Measurement Concept for ABS Appropriated for Microsystem Creation

Sergey Y. Yurish, Nikolay V. Kirianaki, Nestor O. Shpak
Institute of Computer Technologies,
Bandera str., 12, 290013, Lviv, Ukraine
Phone: +380 322 97 16 74; Fax: + 380 322 97 16 41
e-mail: syurish@mail.icmp.lviv.ua

Abstract: The key point of the paper is a reliable ABS solution (smart sensor - encoder - adaptive method of measurement for rotation speed), which is satisfied to all requirements for the modern mid-range (four wheels) and high-end (integrated vehicle dynamics) ABS at lowering the cost of the system. The proposed rotation speed measurement concept is well appropriated for microsystem creation including integrated sensor and metering microcontroller core.

Keywords: ABS, Active Microsensor, Adaptive Measurement.

1. Introduction

Today's cars have become computer networks on wheels. The average car has about $653 worth of electronics and/or electromechanical systems, about 15 percent of the total value of the car [1]. By the year 2000, the average car will contain about $ 950 to $ 2,000 worth of electronics, depending on whether valuation is based on cost or price. It is estimated that electronics will make up 25 to 30 percent of a car's total value [1].
Automotive microsystem applications, especially for anti-lock braking systems (ABS), require high-reliability parts that can withstand the harsh environment and wide temperature range. On the other hand, their price should not be excessively high in conditions of large production volumes.
The reliability of the ABS is determined by speed of processing of measuring information and solution made by the control system. In other words, the ABS must work in the real time. However, in many ABS, the measurement of rotation speed is based on the conventional methods of frequency measurement (standard direct and indirect counting methods) or on the reciprocal (ratiometric) counting method [2]. The choice of measurement technique in the known traditional solutions depends on the desired resolution and data-acquisition rate. For maximum data-acquisition rate, period-measurement techniques are used. Maximum resolution and accuracy are obtained using frequency measurement. On the nature, the time of measurement and conversion for such methods (except the

indirect method of measurement for low frequencies) is redundant in the all specified measuring range of frequencies, except the nominal one.
The key point of the paper is a reliable ABS solution (smart sensor - encoder - adaptive method of measurement for rotation speed), which is satisfied to all the requirements for the modern mid-range (four wheels) and high-end (integrated vehicle dynamics) ABS at lowering the cost of the system (Figure 1.).

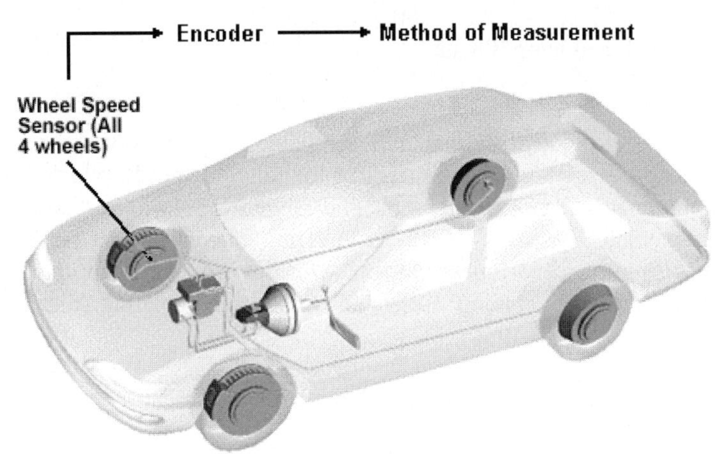

Fig. 1: Three main components of ABS wheel speed sensing concept

2. Smart Wheel Speed Sensor

Sensor technology is playing a critical role in the development of new products and can govern the feasibility of deploying certain systems [3]. The critical sensors involved are those for sensing wheel rotation speed [4].
The active inductive sensor with frequency output was selected as the wheel rotation speed sensor [5]. The rectangular impulse sequence is generated on the sensor's output. Its amplitude is constant and does not depend on the temperatures and direction of wheel rotation. Such signal can be used up to zero speed, without any prior processing at the input (filtration, digitization, etc.). The maximum amplitude of the output signal is determined by voltage supply (in case of the sensor without the input block) or by supply voltage of this input block. The sensor is not influenced by run-out and external magnetic fields in comparison with the usage of Hall sensors. At usage of the Hall sensors it is also necessary to take into account the availability of the initial level of the output signal between electrodes of the Hall's element at absence of magnetic field and its drift. It is especially characteristic for a broad temperature range, defined by road climatic of various countries.

The main performances of the used sensor are:

- 50 % duty cycle;
- Non-contact operation in heavy-duty environment;
- No magnet;
- Any metallic target: steel, copper, brass, aluminium, nickel, iron.

The active inductive micro-sensor [6] is another suitable sensor for microsystem creation. The sensor offers a combination of interesting features, such as position, speed and direction information. It is composed of two silicon chips, one for the integrated micro-coil and others for the integrated interface circuit.
The next step of microsystem creation was integration the microcontroller core with necessary peripherals into system design [7-9]. The instruction set of the microcontroller core was specially optimized for automotive applications. Such modification increases performance with reduced silicon area and permits to include the sensor and other important hardware. The fabrication of such microsystem does not require extra technological steps and can be fabricated by usual CMOS digital processes.

3. Encoder

The usage of the Hall sensors and special encoders in the known ABS reduces to the significant price rise for the system, especially for 4-th wheels. Due to the usage of the active wheel speed inductive sensor the encoder for the proposed sensor microsystem does not require to use the complex construction with oriented magnetic particles. The encoder can be made from the cheaper material, for example, soft steel.
The geometrical teeth sizes of the encoder and one of the possible orientation are shown in Figure 2. Some variations of the geometrical teeth sizes and orientation are possible.

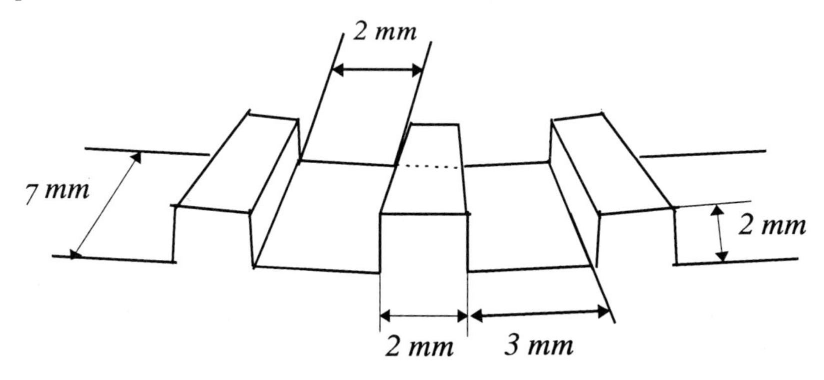

Fig. 2: Teeth geometry for still encoder

4. Adaptive Method of Rotation Speed Measurements

In the proposed approach, based on the so called program-oriented method of dependent count (method with non-redundant reference time interval) [10], the simultaneous synchronous - cyclic frequency measurement of sensor signals is carried out. This frequency is proportional to the rotation speed of wheels. The time of measurement for this method is minimum possible. The quantization error as well as the time of measurement practically depends neither on rotation speed nor exceed beforehand given values. Therefore the sampling rate of the output signal from the sensor ensures the arrival of information in all 4 channels in real time.

Each measuring channel is realised on the virtual level inside the functional-logical architecture of microcontroller.

Let us define this sampling rate for the concrete example. Let automobile be driving with speed of 240 km/h, i.e. 67 m/s, and the time of measurement should not exceed 0.1 sec. (i.e. for this time the automobile will drive no more than 6.7 m at the maximum speed). Let us define the time of quantization having set the relative error of measurement from 0.05 % up to 0.5 % and reference frequency $f_0 = 1/T_0 = 1$ MHz:

$$T = (1 \div 2) \cdot T_0 N = (1 \div 2) \frac{T_0}{\delta}, \tag{1}$$

where N is the minimum number of impulses, which should pass to the counter of the reference frequency according to the necessary relative error δ. This time interval will varied within the limits 2 ÷ 4 ms at $\delta = 0.05$ % and 0.2 ÷ 0.4 ms at $\delta = 0.5$ %. In case, when $f_0 = 10$ MHz this time can be reduced 10 times. Therefore, during 0.1 sec there will be carried out 25 ÷ 50 measurements per second at $\delta = 0.05$ % and 250 ÷ 500 at $\delta = 0.5$ %.

Therefore, the 4 channel frequency-to-code conversion of the ABS control block will execute simultaneous synchronous frequency measurements practically in the real time. So high sampling rate of the signal allows to reduce specific time of measurement up to 0.1 sec without accuracy decrease.

Adaptive possibilities, e. g. automatic choice of the reference time interval depending on the given error of measurement, allow to use the advanced ABS algorithm. Whereas speed is one of the major ABS performances, the required error of measurement can be selected by the microcontroller depending on the current rotation speed due to the adaptive possibilities of the method. It will allow to increase speed at measurement of critical rotation speeds.

The system of measurement will also function successfully at the absence of wheel slip while the rotation speed can be varied in a wide range: from zero speed up to maximum. The essential advantage of the method is the possibility of digital measurement of acceleration with the similar high accuracy and without extra circuitry. It opens the possibility to use the control method not only for

speed, but also for acceleration of rotation as well as combined control method. It opens perspectives to develop the modern high-end ABS for the future needs. For example, road hazards may cause the ABS to function unexpectedly. With the usage of the proposed approach the ABS will make up for road conditions or bad judgment.
Due to minimal possible circuitry necessary for sensor output processing, the metering microcontroller core can be easily embedded into the microsystem.

5. Sensors Network

Sensors network with multilayers architecture is widely used in the systems, with distributed intelligence like a modern car. The advantage of such architecture (Figure.3) consists that high layers use the information ensured with lower layers and do not press in detail of operation the last ones.

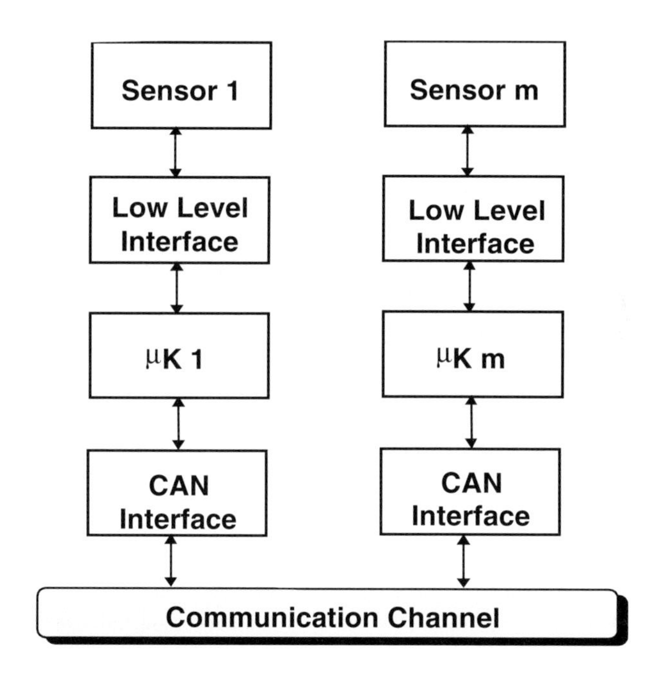

Fig. 3: Multilayers Architecture

The sensor microsystems (one for each wheel) are connected in the network as shown in Figure 3. The proposed microsystem also realises some interfacing functions. Firstly, it is low level hardware and software interface (sensor -

microcontroller), secondly - high level Controller Area Network (CAN) interface, that has been developed for automotive applications to replace the complex cable in cars by a two wire interface [11].

6. Experimental Results

The aim of the experiments was to test the selected sensor element and to determinate the metrological reliability in specified measuring range 0 ÷ 3000 rpm at the specific geometrical sizes of the encoder with 44 steel teeth, to determinate the working air gap and determination of the fitness of the sensor element for the usage in the microsystem for ABS.

The oscillograms of the sensor's output at ~ 3000 rpm are shown in Figure 4. The oscillograms correspond to the sensor operation with the input block. The supply voltage for the sensor was +12 V. The sensor has demonstrated the steady operation at change of supply voltage in the range from + 5.7 up to + 30 V. The measurements were carried out at the working air gap 1.5 mm.

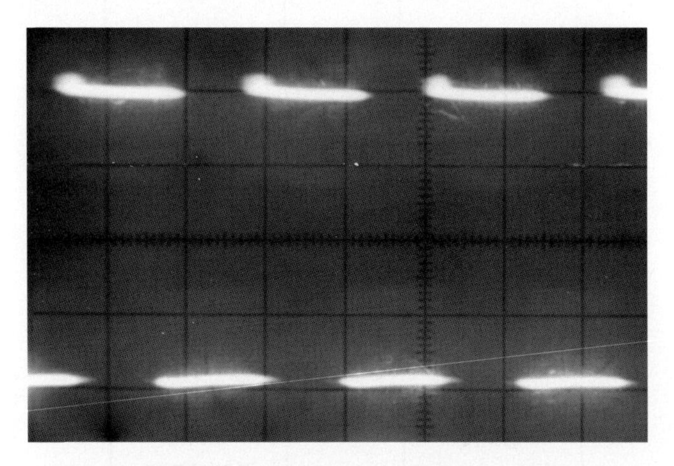

Fig. 4: Oscillogram on the sensor's output at ~ 3000 rpm (1V/div and 0.2 ms/div)

As the rotation speed is connected with the sensor's output frequency by the following dependence:

$$n_x = f_x \cdot \frac{60}{z} = f_x \cdot \frac{60}{44}, \qquad (2)$$

(where z = 44 is the number of teeth), the sensor can be used also for higher rotation speed up to ~ 4090 rpm. Obviously, the sensor can be also used successfully for encoder with 48 teeth. In this case, the maximum rotation speed will be ~ 3750 rpm. The data series for 60 measurement and histogram are shown in Figure 5 a,b.

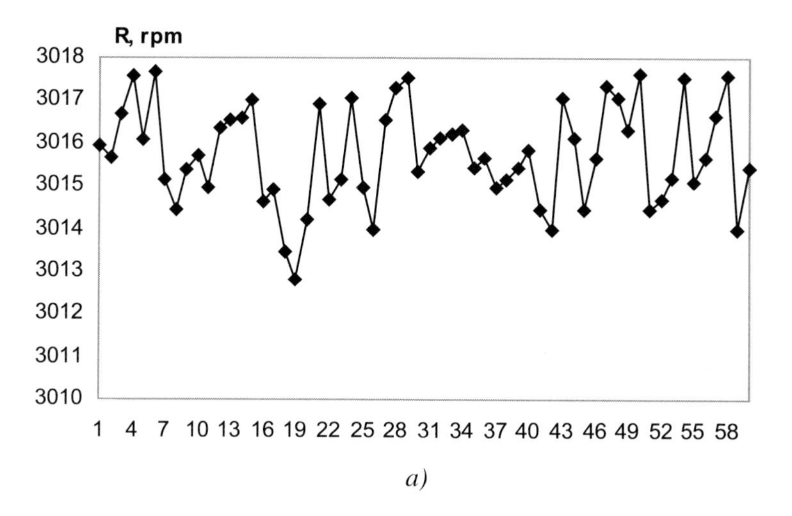

a)

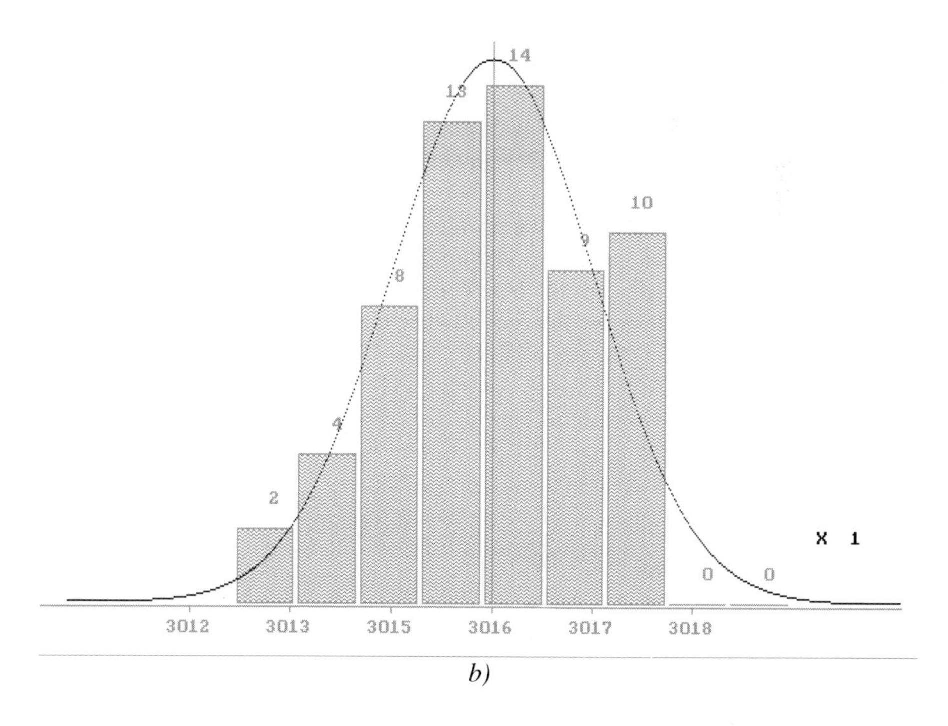

b)

Fig. 5: Data series of measurement (max rotation speed > 3000 rpm)

7. Conclusions

The following advantages of the rotation speed measurement concept are achieved:

- cost reduction;
- more simple and cheap encoders;
- the possibility to be used in all modern wheel and sensor assemblies and realised in many variations;
- economic performance (greater efficiency in the ratio productivity / manufacturability / cost);
- high metrological reliability at miniaturization, weight saving and zero speed capability, and, consequently, increased the car safety;
- the possibility to use advanced ABS algorithm.

Realisation of the rotary sensor together with the metering microcontroller core in the same chip by the usual CMOS digital process will create the basis for new smart microsystems that will have mostly digital signal processing.

References

[1].Car systems integrate more functions (1997, June). J Electronic Components, pp. 186-226.

[2].Burr-Brown Application Handbook (1994), USA,.

[3].Automotive sensors (1998). In: European Sensor News and Technology, September 1998, pp.11-12.

[4].Automotive sensors (1994). Sensors Series, publish by Institute of Physics Publishing, Bristol, UK, ed. M.H. Westrook, J.D. Turner.

[5].Deynega V. P., Kirianaki N. V., Yurish S. Y. (1995). Microcontrollers Compatible Smart Sensor of Rotation Parameters with Frequency Output. In: Proceeding of *21st European Solid State Circuits Conference,* 19-21 September, Lille, FRANCE, 1995, pp.346-349.

[6].MS1020/21 Inductive Micro-Sensor for Position and Speed (1997). Preliminary Technical Data, CSEM Centre Suisse d'Electronique et de Microtechnique SA, Switzerland.

[7].Deynega V.P., Kirianaki N.V., Yurish S.Y. (1997). Intelligent Sensor Microsystem with Microcontroller Core for Rotating Speed Measurements. In: Proceedings of *European Microelectronics Application Conference, Academic Session (EMAC'97),* 28 - 30 May, 1997, Barcelona, SPAIN, pp. 112-115.

[8].Kirianaki N.V., Shpak N.O., Yurish S.Y. (1998). Microcontroller Cores for Metering Applications. In: Proceedings of *International Conference on Programmable Devices and Systems (PDS'98),* 24-25 February, 1998, Gliwice, POLAND, pp. 345-352.

[9].Kirianaki N.V., Shpak N.O., Yurish S.Y. (1998). Modeling of Smart Integrated Sensors of Rotating Parameters for Automotive Application. In: Abstract Booklet of *First International Conference on Modeling and Simulation of Microsystems, Semiconductors, Sensors and Actuators (MSM'98)*, 6-8 April, 1998, Santa Clara, California, USA, T4.2.6.

[10].Kirianaki N.V., Yurish S.Y., Shpak N.O. (1998). New Processing Methods for Microcontrollers Compatible Sensors with Frequency Output. In: Proceedings of the *12th European Conference on Solid-State Transducers and the 9th UK Conference on Sensors and their Applications*, Southampton, UK, 13-16 September 1998, EUROSENSOR XII, ed. N. M. White, Institute of Physics Publishing Bristol and Philadelphia, Sensors Series, vol. 2, pp. 883-886.

[11].Data Transmission Design Seminar (1998), *Reference Manual*, Texas Instruments, USA.

A Novel Technology Platform for Versatile Micromachined Accelerometers

Sebastian Toelg[1], Kathirgamasundram Sooriakumar[2], Yong Hong Loh[2], Uppili Sridhar[3] and Choon How Lau[3]

[1] EG&G Heimann Optoelectronics GmbH,
Wenzel-Jaksch-Str. 31, D-65199 Wiesbaden, Germany
Email: sebastian_toelg@agginc.com

[2] EG&G Singapore Pte Ltd.,
47 Ayer Rajah Crescent, Singapore 139947

[3] Institute of Microelectronics,
10 Science Park Road, Singapore Science Park II, Singapore 117684

Keywords: Accelerometers, Deep Reactive Ion Etching, Lateral Integrated Surface Micromachining

A technology platform called LISA (Lateral Integrated Surface Micromachined Accelerometer) is developed for batch fabrication of differential capacitive accelerometers for a wide range of applications. The novel technology utilizes Deep Reactive Ion Etching (DRIE) of single crystal silicon where the capacitor plates are formed by vertically etching and then releasing the sensing structure by isotropic etch process. The sensing structure is anchored to the substrate by a proprietary oxide isolation technique in which a set of etched trenches are filled with oxide. This novel technology facilitates a high aspect ratio and well controlled plate dimensions.

These key technology developments enabled a platform which allows to fabricate single-axis and dual-axis accelerometer in mass production. Devices are available for wide "g-range" from 5g to 250g. In the near future this will be expanded further down to the 1g-range. The accelerometer consists of two chips for sensor and ASIC containing all signal conditioning. It is packaged either in 16 pin standard SOIC or DIP package. A custom SMT package is available for z-axis applications. Fields of automotive applications include front airbag, side airbag, rollover detection, ABS, active suspension and vehicle dynamics control.

The device has many interesting features including output signal ratiometric with supply voltage, fast power-up time, programmable low-pass filter, patented self-test, high shock resistance, excellent thermal performance and long-term stability. The calibration of accelerometer is done by fusing technology. The design, process and the choice of materials allow the device to meet temperature performance without having to compensate or calibrate for it. This is indeed one of the biggest achievement of this technology platform. Test results demonstrate an offset drift of less than 200mV over temperature range of -40°C to +90°C for 50g device in standard plastic package. Tests performed on higher g-range devices show an offset drift well below 200mV for temperature up to +125°C.

Introduction

Over the last years electronics had a tremendous impact on our automobiles. In the beginning electronic subsystems replaced existing mechanical or hydraulic solutions serving the same functions but with better performance, higher reliability, more compact dimensions, less weight and finally lower price. Well-known examples are electronic fuel injection systems and electronic control of automatic gear boxes.

Electronics once being introduced offered increased functionality by combining existing separate subsystems to perform higher level functions in an integrated system. For example the electronic fuel injection, electronic ignition system, misfire detection, automatic gear control, exhaust fume management, etc. have been integrated to a comprehensive power train management system. The result is a highly improved performance with more powerful engines, lower gas consumption, less emission to the environment and higher reliability.

The third generation of electronic systems is characterized by facilitating totally new functions that could not be achieved otherwise. Such systems generally rely on sophisticated models of the vehicle or components thereof. The complex algorithms generally require powerful μ-controllers or DSPs. But the smartest algorithm and the fastest processor would be idle if there was no access to what is going on in the reality of the outside world. This is the reason for the strong and rapidly increasing demand for advanced sensors in automotive applications.

Highly sophisticated sensors are the interface of the electronics to the real world. The requirements on automotive sensors for pressure, torque, position, velocity, acceleration, mass flow, etc. are extremely demanding in terms of performance, reliability, size, weight, temperature range, hostile environmental conditions, lifetime, etc.

In the sequel, the focus will be on sensors for acceleration, i.e. accelerometers. The automotive applications for accelerometers are numerous. Maybe most attention in public is paid to restraint systems or safety systems like airbag and belt pretensioner. Here, dedicated sensors for detection of front and side impact are required. Recently, the airbag concept has been expanded to window curtains that require accelerometers with much higher sensitivity in combination with angular rate sensors for rollover detection. Other applications for accelerometers are much less obvious and well-known. Examples are anti-lock brakes (ABS), traction control, active and adaptive suspension, active roll stabilization, vehicle dynamics control (VDC) or electronic stability program (ESP) and rough road detection in conjunction with misfire detection.

Accelerometer Technologies

Accelerometers generally make use of a simple measurement principle and consist of a few constituent parts. A seismic mass or proof mass experiences a force due to external acceleration of the sensor device. This mass is somehow suspended on a spring. Within the limits of Hook's law the force results in a displacement of the proof mass that is proportional to the applied acceleration. This displacement is converted to an electrical signal. This is where different physical means for sensing come into play that give rise to various sensor designs and manufacturing technologies that are reflected in subsequent generation of accelerometers. The following table gives an overview.

Generation	Technology	Sensing	DC Response	Self-test	Durability	Cost
1	Electro-Mechanic	various	yes	no	o	-
2	Quartz	piezo-electric	no	some	o	-
3	Si bulk MEMS	piezo-resistive	yes	some	+	o
4	Poly-Si surface MEMS	capacitive	yes	yes	o	+
5	Single-Si surface MEMS	capacitive	yes	yes	+	+

In principle, crash detectors and accelerometers can of course be manufactured using conventional mechanical engineering. However, the strong demand for cost reduction, high volume, performance, miniaturization, long-term reliability could only be met by developing and introducing MEMS (Micro Electro-Mechanical

Systems) technology. This enables high volume production using bulk manufacturing of hundreds of sensors at a time with little tolerance.

Bulk silicon micromachined accelerometers have been designed for general purpose applications in industrial, medical, aerospace and consumer applications. These accelerometers can offer very good performance, but usually do not meet the cost targets of today's and future automotive and high volume consumer applications. A bulk micromachined accelerometer is a multi-layer device (usually 3 layers). Each layer consists of a silicon wafer that is processed on both sides or at least on one side. The thickness of the wafer is limited due to etch time to form the sensing structure. Thicker wafers have also a negative influence on tight control over the process parameters. This makes it more difficult to maintain the required tolerance. The mechanical stability limits the wafer size that can be handled to 4" in most cases. Acceleration is sensed by stain induced change of conductivity in resistors that are implanted in the suspension beams of the seismic mass. Unfortunately, these resistors exhibit a significant temperature dependence that has to be compensated and calibrated. The sensing direction for acceleration is usually perpendicular to the wafer plane.

The advantages of surface silicon micromachining as compared to bulk micromachining are low production cost, compactness of sensor and potential integration with circuitry for signal conditioning. For a number of years the ability to form sensing structures in poly-silicon has been commercialized. Acceleration is usually sensed by measuring the change in capacitance due to the lateral displacement, i.e. in the wafer plane, of the capacitor plates formed by the seismic mass. The manufacturing processes are quite similar to standard processes used for integrated circuits. First, a sacrificial layer (usually oxide) is deposited on the single crystal silicon substrate. Subsequently a thin poly-silicon layer is deposited on top. By means of patterned photo resist, the sensing structure is etched out of the poly-silicon. The sacrificial layer serves as etch stop. In the next step, the sacrificial layer is removed leaving the free sensing structure attached to special anchor points only. These anchor points can be formed by holes patterned into the sacrificial layer before depositing the poly-silicon. Both layers, sacrificial and poly-silicon, are only a few microns thick, typically 1-2μm. The signal level of this type of sensors is typically quite low due to small seismic mass and small differential capacitance. Moreover, the small free space (given by the thickness of the sacrificial layer) under the sensing structure and the substrate results in a parasitic capacitance. Taken all these technological difficulties together, the sensor structure and the circuitry for amplification and signal conditioning has to be integrated on the same chip. Additional parasitic capacitance caused by wires connecting two separate chips (sensor and ASIC) would swamp the small sensor signal. The deposition processes give rise to additional tolerances and the poly-silicon does by far not meet the superior material properties of single crystal silicon. The complexity and the limitations of the processes involved decreases the achievable yield and adds cost to the product.

The process steps for depositing various layers are avoided by LISA technology. Here the sensing structure is directly etched out of single crystal silicon of the substrate using advanced deep reactive ion etching technology (DRIE). The mechanical, thermal and long-term stability of the sensing structure greatly benefits from the superior material properties of single crystal silicon. In addition the thickness of the sensing structure is 5-10 times larger than the structures that can be achieved with well-known poly-silicon processes. The resulting high proof mass and large capacitance lead to a strong sensor signal that can be easily connected to a separate ASIC through wires. The number of process steps is significantly smaller and the manufacturing is less complicated. As a result, the sensors using this technology provide better performance and are less expensive.

Concept

The strong signal of the acceleration sensor manufactured in LISA technology allows the circuitry for signal conditioning to be on a separate chip. The sensor chip is connected to the ASIC using standard wire bonding technique to form an MCM. This two-chip approach has several advantages over integrated solutions:

- faster development time because of reduced complexity
- clearly defined interface between sensor element and electronic circuitry
- same ASIC can be combined with several sensor designs, e.g. for different g-ranges
- same sensor element can be combined with different ASICs, e.g. for analog and digital output
- ASIC can be manufactured using well established standard IC processes
- manufacturing process for sensor is much simpler and only few masks are needed
- processes can be better optimized for cost
- the total yield can be higher because of less complicated processes and intermediate check points for sensor element and ASIC before assembly

The displacement of the proof mass due to applied acceleration is picked up by a differential capacitive structure. A set of moving plates in between two other sets of fixed plates form the electrodes of two capacitors (Fig. 1). The moving plates are attached to the proof mass of the sensing element. As the proof mass is deflected due to acceleration, the middle set of plates moves accordingly. The distance to one set of fixed plates increases while the distance to the other fixed set of plates decreases. Hence, the capacitance of the first pair of plates decreases while the capacitance of the second pair increases, respectively. This change of capacitance is picked-up by the electronics.

Sensor Element

Silicon etching is becoming recognized as an enabling technology for Micro Electro-Mechanical Systems (MEMS). LISA technology is based on surface micromachining technology of single crystal silicon where the sensing element is fabricated with proprietary processes. Three major process steps are described here:

- **Trench etch:** Optimized deep reactive ion etching (DRIE) is used to achieve profile anisotropy independent of crystal orientation. Our deep dry trench etch technology using a plasma etcher renders high aspect ratios possible - typical in the order of 1:10, but much higher ratios are feasible. Some advantages of the applied advanced etch processes are: high etch rates, high selectivity to mask, highly anisotropic and well controlled side wall profile resulting in high aspect ratio capability. The width of the trenches and the plates is typically about 1-2μm (Fig. 2). The trenches are usually 20-30μm deep for high volume production, but much deeper structures are possible. The angle of the smooth side walls is controlled within a tolerance of less than 1°.

- **Release etch:** The walls of the plates are passivated for protection and the silicon at the bottom of the trench is exposed to reactive ion etching. The process parameters are modified and optimized for release etch. In this release etch process step the structure of plates is detached from the substrate and the sensing structure is free to move (Fig. 3). The free space between released structure and substrate is typically 10-20μm. This leads to a low parasitic capacitance.

- **Trench oxide isolation:** A new technology called trench oxide isolation (TOI) is used to electrically isolate the whole sensor structure from the substrate. The trenches for isolation comb structures are etched using DRIE technology. Subsequently, the trenches between the silicon side walls are refilled with silicon oxide. This proprietary process relies on an optimized combination of low pressure chemical vapor deposited silicon oxide (LPCVD) and thermal oxidation of silicon. This technique works well for trench widths of 1μm to 3μm. The silicon fingers of the supporting comb structure have about the same width. The sensor structure is anchored on three sides and the bottom by oxide. The big advantages are complete electrical isolation from the substrate and low parasitic capacitance. At the same time the oxide filled comb structure gives very stable mechanical anchor points with excellent thermal and long-term behavior.

Fig. 4 shows the sensor element with fixed and moving capacitor plates, TOI anchor points and spring with end stop for over range protection. Fig. 5 depicts the interlaced fingers of fixed and moving capacitor plates. The high aspect ratio and the excellent surface of the side walls is obvious.

The unique deep trench etching capability of LISA technology renders possible to manufacture a single crystal sensor element with high aspect ratio. Single crystal silicon exhibits excellent material properties and is one of the most well-known materials in the world. This inherently provides a sensor element with excellent linearity, which is insensitive to temperature, exhibits excellent long-term and gives strong signal.

The high aspect ratio of the sensing structure results in a large proof mass and enables a larger capacitance to be produced on the same area of silicon compared to other processes using poly-silicon. One of the advantages is that the sensing element and the circuitry (ASIC) can be separated on two chips. Thus, the high complexity of manufacturing an integrated device (single chip) in order to minimize parasitic capacitance as required by some poly-silicon technologies can be avoided. Moreover, the large thickness of the structure drastically increases the stiffness of the structure normal to the wafer plane and results in excellent shock resistance in this direction.

The sensor structure is fabricated using only a few mask steps with standard semiconductor equipment. Sensor elements are fabricated on 6" wafers. A cap wafer is bonded on the processed sensor wafer to protect the sensor structures. This is done before sawing and dicing the sensor elements. Hence, each sensor element is hermetically sealed to avoid contamination during further fabrication. Using a cap made of the same material as the sensor wafer results in excellent thermal stability. The applied bonding technique reduces induced stress to a minimum.

ASIC

The ASIC circuitry consists of several building blocks (Fig. 6). The sensor element is connected to the position amplifier that converts the change in capacitance to a change in voltage thus being proportional to the exerted acceleration. Due to the high capacitance change ratio of the sensing element, the ASIC is able to detect displacements of the proof mass well below 1nm. The voltage signal passes the calibration stage to adjust for gain and offset according to the parameter programmed into the fuses. An on-chip low-pass filter can be programmed at 500Hz or 2kHz cut-off frequency at the factory per customer request. This filter is realized in switched capacitor technology. A subsequent continuous time filter removes some low magnitude artifacts caused by the previous discreet time signal processing. Additional building blocks are: oscillator and clock generation, programmable fuse circuitry and control of self-test.

The ASIC for signal conditioning (amplification, filtering, calibration, etc.) and self-test control is fabricated in a well established 0.8μm double poly, double metal standard CMOS process.

The device is operational for supply voltage between 3V and 7V. The output signal is ratiometric with the supply voltage. The output voltage swing is from 0.5V for minimum acceleration and 4.5V for maximum acceleration under nominal 5V supply condition. The offset is usually calibrated to give 2.5V output for zero acceleration. However, due to versatile ASIC design the offset can be shifted over a large range to allow even for highly asymmetric g-range, e.g. –50g to +250g. Note, this can be done by programming the ASIC only, i.e. without modification of the sensor element.

The self-test is activated by a designated CMOS-logic input. A logic high level applied to this pin, causes a predefined electrostatic force exerted on the sensing element. This force deflects the proof mass and the induced displacement simulates an external acceleration and results in a corresponding output signal. Thus, during activated self-test the sensor element as well as the whole signal path can be verified to be fully functioning. During normal operation this self-test pin is at a logic low level (grounded).

Packaging and Different Accelerometer Versions

The hermetically sealed sensor element is packaged together with the ASIC. The wires between sensor element and ASIC are bonded together with the connections of the ASIC to the external pins. Thanks to the hermetically sealed sensor element and its excellent stability and thermal performance the accelerometer can be packaged in standard injection molded plastic package. The single-axis sensor with sensing direction in the PCB plane are available in 16 pin wide body SOIC package for SMT mounting or in 16 pin DIP. The standard plastic packages are available in high volume and offer low cost, small size, high reliability, mounting flexibility and easy assembly using automatic pick-and-place tools.

Only four pins have to be connected by the end user: 5V power supply, ground, signal output and self-test activation input (Fig. 7).

A dual-axis accelerometer (Fig. 8) with the same performance and electrical properties is also available. Dual-axis accelerometers are well suited for more advanced passenger protection where one axis senses front impact and the perpendicular axis senses side impact. Here, the sensor chip comprises two separate and identical sensing structures rotated by 90° with respect to each other. Thus, the two output signals correspond to the acceleration in the x and y direction of the chip plane. The ASIC comprises two separate functional signal paths each devoted to one channel. The calibration for offset and sensitivity can be adjusted independently for each channel. Single and dual-axis versions are available in the same package (16 pin wide body SOIC) and have compatible pin layout (Fig. 7) to facilitate easy exchangeability. The dual-axis accelerometer uses only one additional pin (#12) for the second output signal. Both channels share a common input for self-test activation. Compatible pin layout of single and dual-axis versions enables the use of common PCB for simpler front impact ECU and for more advanced ECU offering front and side impact protection.

The third version of LISA accelerometers is a z-axis sensor where the sensitive axis is perpendicular to the mounting plane of the PCB. This geometry is particularly well suited for side impact satellites that are mounted in the B-pillars or in the doors. Here, space limitations are generally a concern. This geometry is achieved by means of a custom plastic SMT package where the chips are mounted sideways. This low profile package has a 16 pin narrow body SOIC footprint and is pin compatible (Fig. 7) with the standard single-axis version in SOIC package.

Calibration

In order to compensate for unavoidable variations during manufacturing and to improve yield it is common practice to test and trim every part at the factory prior to shipment to the end-user. One approach for calibration is by means of laser trimming either of the mechanical sensor structure or of thin-film resistors. This has to be done on the accessible and therefore open chips, i.e. before final packaging. However, being a mechanical device, the accelerometer is inevitably sensitive to packaging stress.

To avoid stress problems the test and calibration procedures, e.g. for offset and sensitivity, should be carried out after final package assembly. This requires some kind of non-volatile programmable on-chip memory. Two forms of such programmable memory are well established - using either EPROM or fuse technology. EPROMs require additional processing steps and costs. The fuse technology applied here can be fabricated easily in standard CMOS processes. Fuse technology is also more reliable over the automotive live cycle and temperature range. Reliability and robustness is further enhanced by adopting differential fuse circuit topology.

Final calibration and testing of the packaged parts is performed on a shaker. For verification and documentation the content of the fuses is read out after programming. A serial communication interface is built into the fuse circuitry for communication with the external world during programming and read-out. For the dual-axis accelerometer, the I/O pins for serial communication are shared among both channels. The end-user does not need to care about these pins.

Summary, Performance and Conclusion

Based on the LISA technology a comprehensive and highly compatible product family has been designed that covers the whole range of safety critical automotive applications from single and dual-axis as well as z-axis accelerometers. It offers significant advantage in price and performance compared to conventional technologies.

Current LISA accelerometers are two-chip designs. A rugged sensor chip and an ASIC are combined in a 16 pin wide body SOIC package to allow for small size, low cost, high reliability and mounting flexibility. The accelerometer consists of a surface silicon micromachined sensing element and an ASIC. Due to acceleration, the sensing structure moves and changes the differential capacitor. It is fabricated from single crystal silicon with a proprietary process. The etch profile with a high aspect ratio improves signal to noise performance of the sensor compared to poly-Si structures. The single crystal silicon inherently provides a sensor element, which is insensitive to temperature and exhibits excellent long-term stability. The ASIC for signal conditioning is fabricated with a 0.8 μm double poly, double metal CMOS process.

The accelerometer exhibits a high shock resistance and provides a patented self-test feature that produces an output signal proportional to a simulated acceleration. The device has terminals for supply voltage, self-test input and device output signal. Each accelerometer is individually serialized. Sensor stability and performance has eliminated the need for temperature compensation resulting in improved performance and reduced cost. The advanced technologies applied to sensor element and ASIC allow the accelerometer to be packaged as a 16 pin wide body SOIC.

In plastic package, the accelerometer with ±50g demonstrates a non-linearity of less than 0.1%. The ratiometric error is below 2% for supply voltage between 4.75V and 5.25V. The offset remains between 2.3V and 2.7V throughout the whole standard operating range of –40°C to +90°C.

LISA accelerometers provide excellent performance at low price to meet the challenging requirements of wide range of automotive, industrial and consumer applications. LISA technology is well suited to continuously provide high quality products and best solution especially for safety critical applications at very competitive price. Based on LISA technology we will continuously develop further inertial sensors like high resolution low-g and angular rate sensors for various applications.

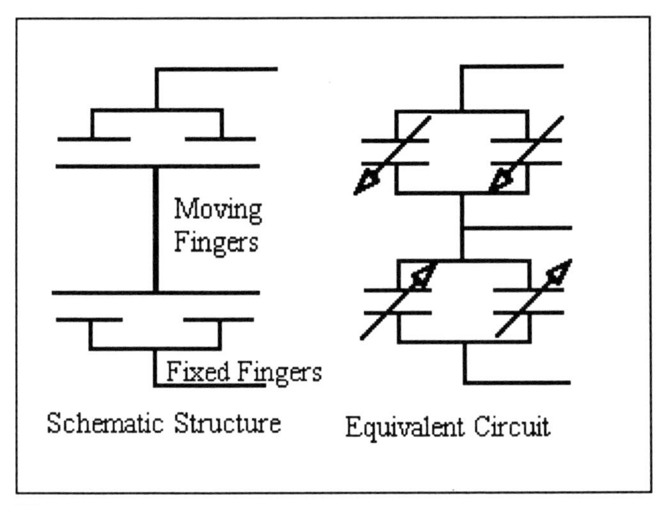

Fig. 1

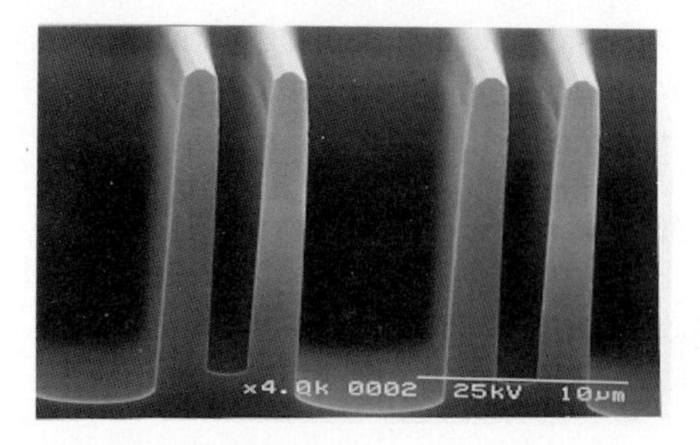

Fig. 2 Trench Etch

Fig. 3 Release Etch

Fig. 4 Sensing Structure

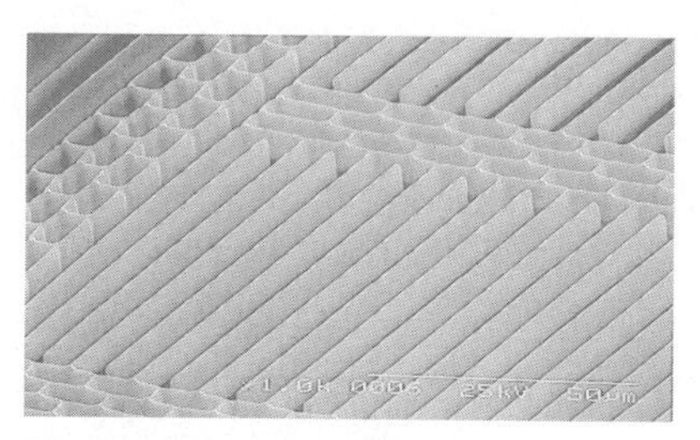

Fig. 5 Capacitive Plates

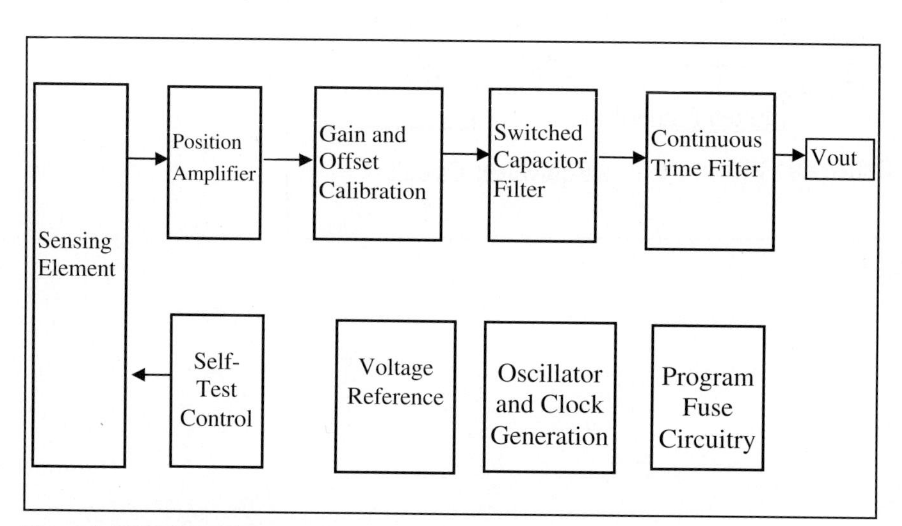

Fig. 6 ASIC Block Diagramm

Signal	Pin	Pin	Signal
NC	1	16	Vdd
NC	2	15	GND
NC	3	14	Y out (Z out)
NC	4	13	Self-Test
NC	5	12	X out (Single Axis = NC)
NC	6	11	NC
NC	7	10	NC
NC	8	9	NC

Fig. 7 Pin Layout

Fig. 8 Accelerometer in SOIC Package

A Low Cost Angular Rate Sensor for Automotive Applications in Surface Micromachining Technology

R. Schellin, A. Thomae, M. Lang, W. Bauer, J. Mohaupt, G. Bischopink, L. Tanten, H. Baumann, H. Emmerich, S. Pintér, J. Marek, K. Funk*, G. Lorenz*, R. Neul*

Robert Bosch GmbH, Departments K8/STZ, K8/EIC, Tübinger Str. 123, 72703 Reutlingen;
* Robert Bosch GmbH, Department FV/FLI, P.O Box 106050, 70049 Stuttgart

Abstract: A second generation angular rate sensor suitable for high volume production is presented. The sensor consists of an electrostatically driven oscillating disk, which reacts – due to the conservation law for angular momentums - with a tilt movement when an angular rate is applied. This tilt movement is detected capacitively by electrodes on the substrate underneath the oscillating disk. The sensor element is manufactured by using a standard surface micromachining process. The signal evaluation is realised in a switched capacitor technique. The PLCC44-packaged sensor provides a self test function and may be calibrated to a customer specific sensitivity (currently 7 mV/°/sec for a +/-250 °/sec Full Scale measurement range). The sensor is designed for automotive applications, especially for rollover sensing.

Keywords: Angular Rate Sensor, Surface Micromachining, Rollover Sensing

1. Introduction

During the last years micromachining gained more and more importance and influence in different, especially industrial applications. The automotive sector is one of the most innovative and most rapidly growing markets for different kinds of sensors [1]. Beyond the need for information about pressure and acceleration, the measurement of an angular rate with high resolution is one of the most important aims in the development of automotive comfort and safety systems (see, e. g. [2-6]). Especially, rollover sensing and navigation control are fields of high interest. In contrast to other kinds of sensors, a gyroscope always consists of an active part, which has to be driven mechanically in a vibrating movement, and a non-driven part, which reacts to an applied angular rate. The sensor presented here contains a rotationally oscillating disk as acting part and two fixed electrodes, forming – in combination with the oscillating disk – the sensing part.

2. Micromechanical sensor element

2. 1. Design and functional principle

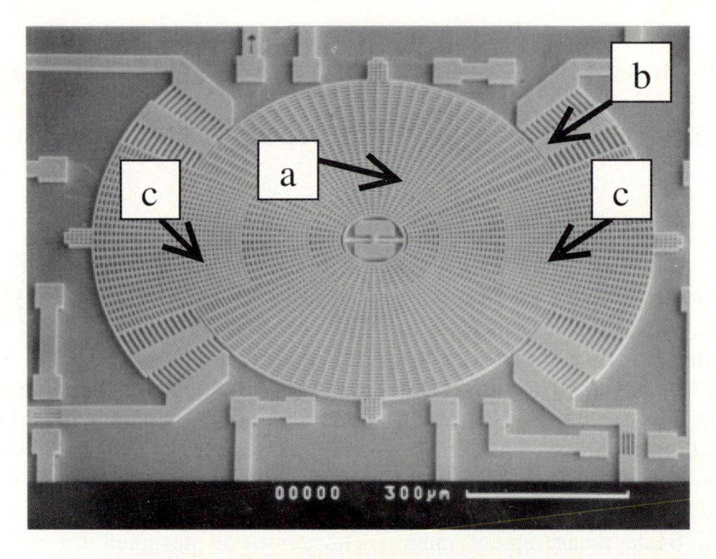

Fig. 1: SEM-picture of the sensor element
(a = sensing mass, b = interdigitated comb structure,
c = detection electrodes – unvisible - underneath oscillating disk)

A SEM-picture of the sensor element is shown in Fig. 1. It consists of an oscillating disk *(a)*, which is driven in-plane (rotational oscillation: f_z approx. 1500 Hz) by applying an electrostatic voltage across interdigitated comb structures *(b)*. An angular rate Ω_y around an axis in the chip plane produces a tilt moment M around the orthogonal in-plane axis (Fig. 2)

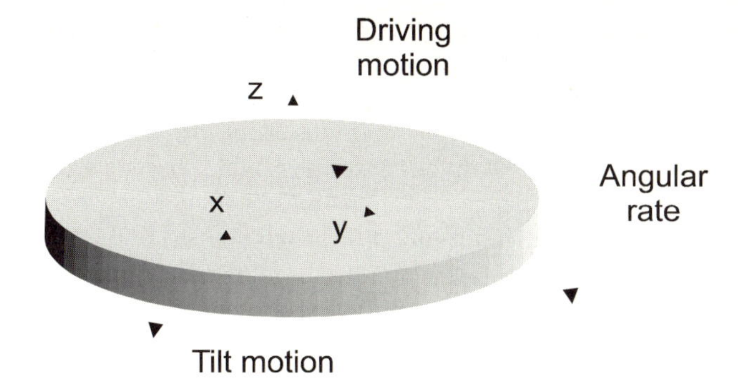

Fig. 2: Function principle of the sensor element

and is thus transformed into a torsional out-of-plane deflection of the center mass (see Fig. 3a).

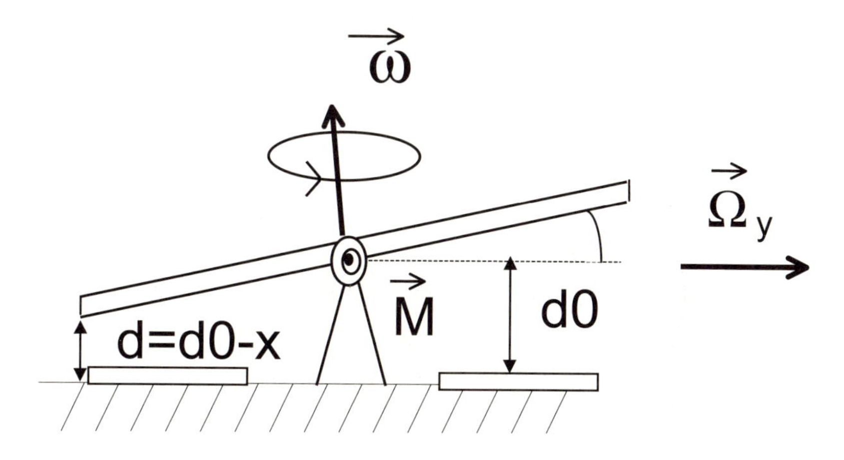

Fig. 3a: Tilt angle resulting from angular momentum on rotating disk

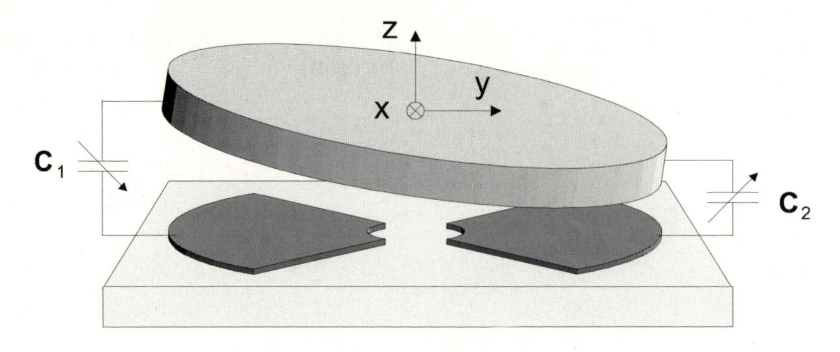

Fig. 3b: Capacitance change resulting from angular momentum on rotating disk

This deflection x is capacitively detected by using a differential capacitor structure (the two counter electrodes are placed underneath the oscillating mass, see Fig. 3b). The signal evaluation of the sensor element is realised in a switched capacitor technique (see section 3). Thus, the changes in the capacitances are transformed into an output voltage, which is proportional to the angular rate.

2. 2 Technology

Prototypes of the sensor element are fabricated by using the surface micromachining technology of the Robert Bosch GmbH. The main process steps are briefly described below, more details are given in [7,8].

On top of a 6“ silicon wafer an insulating SiO_2-layer is thermally grown. In the next step a LPCVD-polysilicon-layer is deposited and structured to be used as conductors and lower electrodes (Fig. 4a).

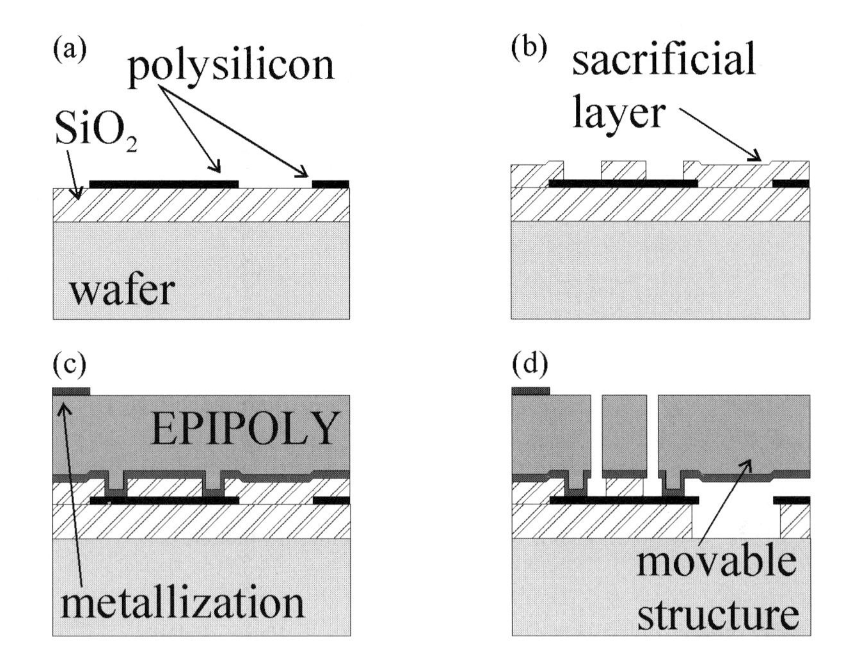

Fig. 4: Process flow of the sensor

Then a second oxide, the sacrificial layer, is deposited and structured (Fig. 4 b) in order to form contact holes to the buried electrodes. This step is followed by the deposition of the upper LPCVD-polysilicon, which serves as a starting layer for the epitaxial deposition of a 11 µm polysilicon layer (so-called Epipoly, EP). The mechanical parameters (especially internal stress and stress gradients) of this Epipoly are optimised by choosing special doping and annealing conditions. Subsequently, aluminum is sputtered and patterned to form bond pads (Fig. 4 c and Fig. 5). Then, the Epipoly is structured with aspect ratios up to 5:1 in depth. Afterwards, the sacrificial layer is etched with vaporised HF (Fig. 4 d). Finally, the whole sensor structure is encapsulated with a silicon cap, which is placed on top of an Epipoly frame surrounding the sensor structure itself (Fig. 5).

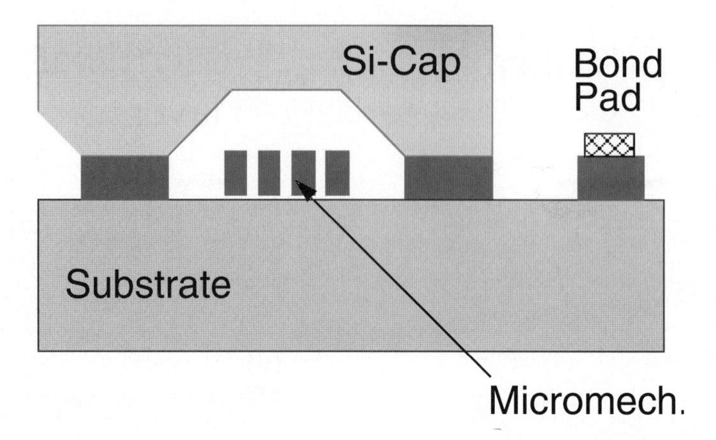

Fig. 5: Sensor element with silicon-encapsulation

In order to obtain a high sensitivity, high quality factors are needed (see section 2.3). Thus, a very low pressure close to vacuum inside the encapsulated volume is necessary. This is achieved by bonding the Si-micromachined cap under vacuum conditions (less than 5 mbar). The whole bonding procedure is performed on wafer level.

2.3 Theoretical Considerations

The complex sensor movement (in case of an applied angular rate Ω_y) can be described with the two following coupled differential equations:

$$0 = (J_x + J_z - J_y)\Omega_y \dot{\gamma} + (J_z - J_y)\dot{\beta}\dot{\gamma} + J_x \ddot{\alpha} + d_x \dot{\alpha} + k_x \alpha \qquad (1)$$

$$0 = (J_x - J_z)\dot{\alpha}\dot{\gamma} + J_y \ddot{\beta} + d_y \dot{\beta} + k_y \beta \qquad (2)$$

with the simplified solution for the tilt angle α (sensing mode)

$$\alpha = \frac{2}{\omega_z} \frac{\gamma_0 \Omega_y}{\sqrt{\left(\frac{\omega_x^2}{\omega_z^2} - 1\right)^2 + \frac{\omega_x^2}{Q_x^2 \omega_z^2}}} \qquad (3)$$

where d_x and d_y are the damping coefficients, k_x and k_y the torsional spring constants, ω_z and ω_x the resonance frequencies of the driving and the detection mode, γ_0 the amplitude of the in-plane oscillation, Q_x the quality factor of the sensing mode, J_x, J_y, and J_z the moments of inertia, β the tilt angle (y-axis) and Ω_y the angular rate. The moments of inertia can be transformed into a factor 2 due to the specific aspect ratio of the sensor structure.

The tilt angle α and thus the sensitivity is mainly dependent on the relation between the driving frequency ω_z and the detection resonance frequency ω_x. In order to obtain a high sensitivity and a low temperature dependence of the sensor output signal, especially the resonance frequencies mentioned above have to be optimised. For quality factors higher than 10, its influence on the sensitivity can be neglected. In addition to the quality factor of the sensing mode, the driving mode also requires low pressure in order to enable an in-plane oscillation with reasonable driving voltages.

More details about the sensor modelling are given in [9].

2.4 Packaging

As sensor package a standard PLCC44 housing is used. In order to minimise any mechanical stress inside the module, the micromechanical sensor element and the evaluation circuit are mounted on a dedicated lead frame with special low stress glue. The two chips are connected together by chip-to-chip-bonds.

3. Electronic Circuit

The control circuit of the sensor element can be separated into two parts: the driving and the sensing paths. Fig. 6 shows a block diagram of the signal processing with its two coupled paths.

3.1 Driving Path

The sensor structure is driven by electrostatic forces at one pair of interdigitated combs. The other pair of combs is used to detect the in-plane deflection of the element. The evaluation of the capacitor value is realised by SC-based C/V-converters which work with a carrier frequency of about 24 kHz (HF). As shown in equation 3 the sensitivity of the sensor is proportional to the amplitude of the in-plane oscillation. To guarantee a constant in-plane deflection over temperature and other parameters a regulation of the amplitude with an automatic gain control (AGC) is necessary.

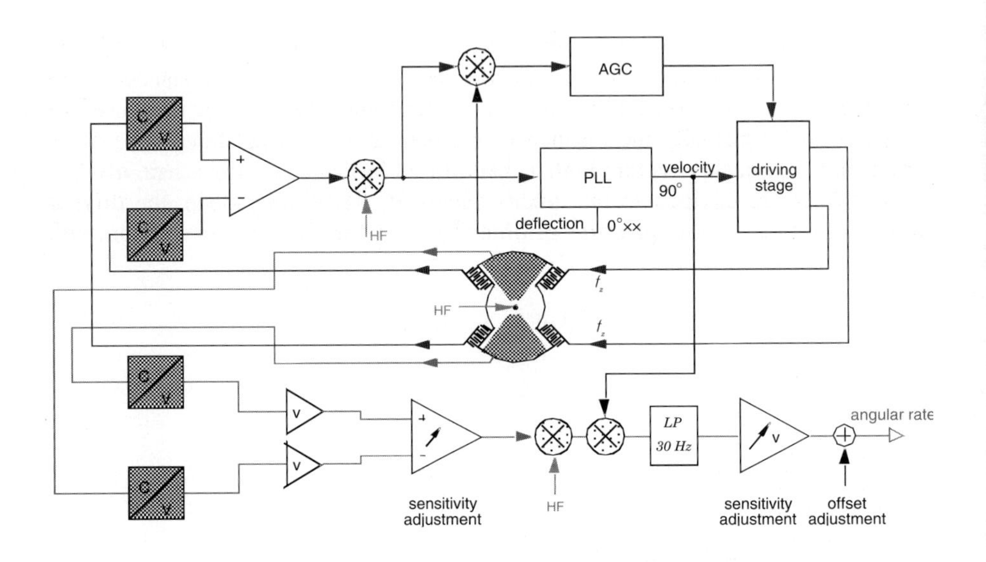

Fig. 6: Block diagram of the sensor signal processing

The deattenuation of the mechanical element with its high quality factor requires a high amplification as well as a 90 degree phase shift. The latter is accomplished by a PLL (phase locked loop). The PLL is synchronised to a signal which is in phase with the in-plane deflection of the sensor element and provides a 90 degree phase shifted signal which is in phase with the in-plane velocity.

3.2 Detecting Path

The deflection 'out-of-plane' caused by the external angular rate leads to a change of the 'out-of-plane capacitances'. This change is converted into a voltage by SC-based C/V-converters in the same way as in the driving path. The signals are amplified and after the differential-to-single-ended conversion, a demodulation with the HF is performed. Synchronous demodulation with the resonance frequency (in phase to the in-plane velocity) provides the information of the angular rate in the base band. The low pass filter determines the bandwidth of the system which can be externally tuned by the customer. Variations of sensitivity and offset of the micromechanical sensor element are adjusted in the output stage.

4. Characterisation

4.1 Transfer function of prototype sensor element

Fig. 7 shows the characteristic transfer functions ('out-of-plane' and 'in-plane' mode) of a prototype gyroscope (the resonance frequencies of the final micromechanical sensor elements are slightly different from those in Fig. 7). The resonance frequency f_z of the in-plane mode is 1.35 kHz, whereas the resonance frequency of the out-of-plane mode is slightly higher (1.75 kHz). As expected, the quality factors are significantly higher than 100 for both vibration modes. This is due to the low pressure (<5 mbar) inside the sensor element.

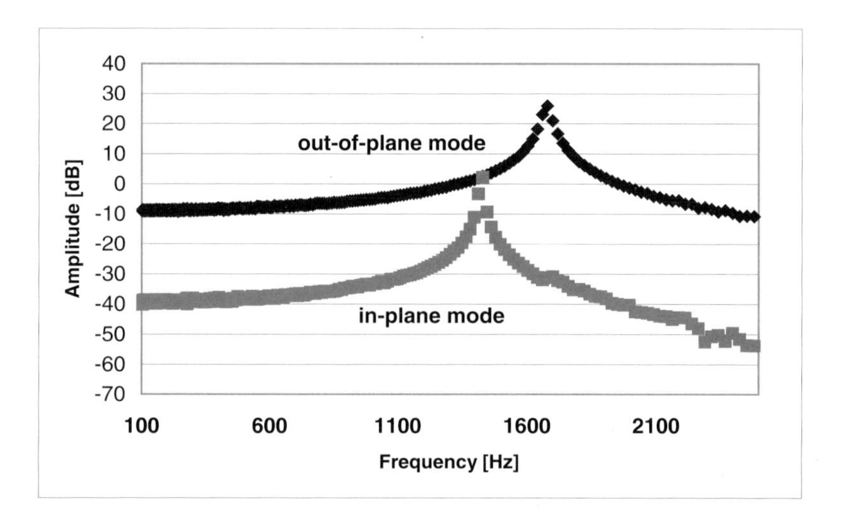

Fig. 7: Transfer characteristics of the in-plane and out-of-plane mode of a prototype sensor element

4.2 Output characteristic

The output characteristic of an angular rate sensor (at room temperature) is shown in Fig. 8. This characteristic was measured by synchronous demodulation of the sensor element output signal with the in-plane-drive signal (see section 3). The sensor shows a good linearity in the whole measurement range. The final sensitivity was obtained by calibrating the sensor to approx. 7 mV/°/sec. This ensures a Full Scale measurement range of +/-250 °/sec for a 5 V power supply.

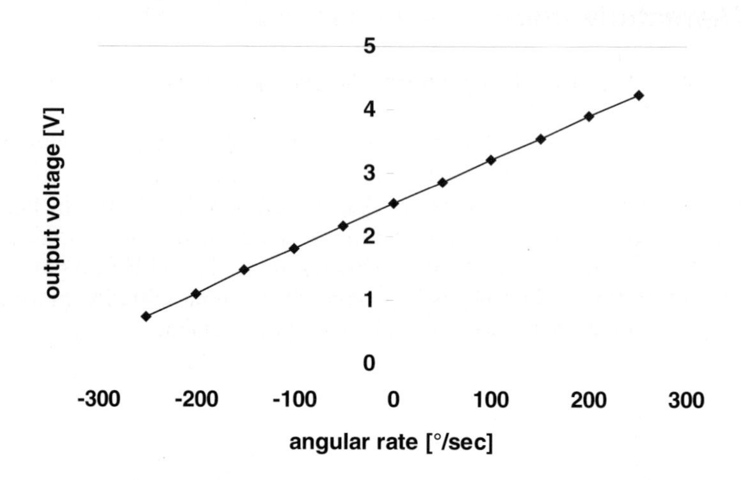

Fig. 8: Output characteristic at room temperature

4.3 Temperature dependence of offset

The temperature dependence of the offset voltage ($U_{OFF} = U_{OUT}$ - 2.5 V) of the sensor without applying an angular rate (see Fig. 9) leads to a total change in the output voltage of 0.6 % (relative to 2.5 V) in the temperature range [-40 °C .. +85 °C].

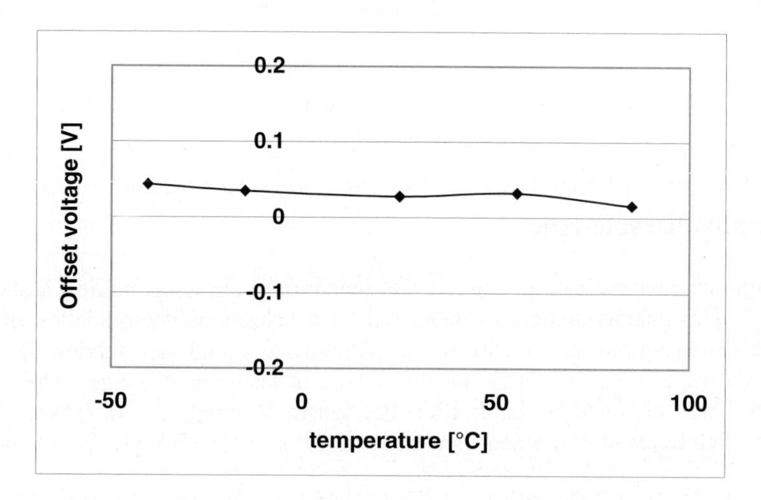

Fig. 9: Temperature dependence of zero-degree output voltage

4.4 Temperature dependence of sensitivity

The temperature dependence of the sensitivity (in % relative to the room temperature sensitivity) of a sensor is shown in Fig. 10

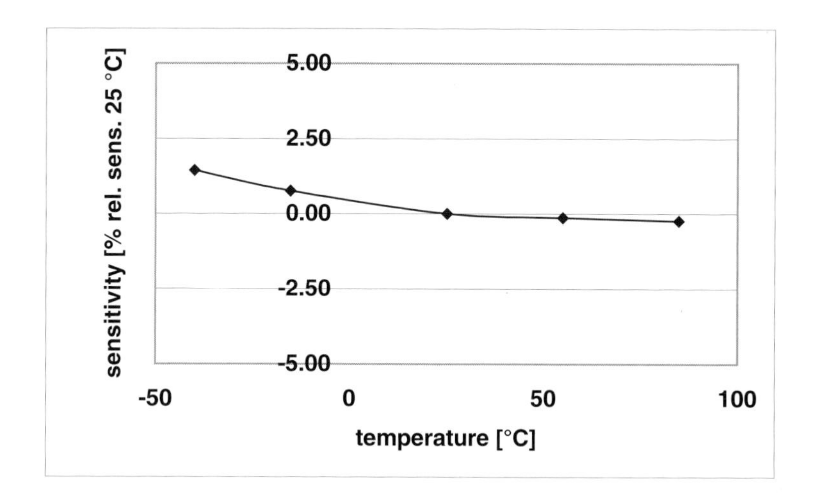

Fig. 10: Temperature dependence of sensitivity (in % relative to the room temperature sensitivity)

The sensitivity changes less than 2 % in the temperature range –40 °C .. +85 °C.

4.5 Linearity

The non-linearity of the sensor is lower than 1 % over the whole Full Scale measurement range, which is sufficient for most of the applications, especially for rollover sensing.

4.6 Noise

The minimal detectable angular rate of the sensors is typically less than 2 °/sec. This is sufficient for rollover sensing, but still too high for navigation control. Improvements can be achieved, especially by optimising the noise behaviour of the first amplification stage and by increasing the sensitivity of the sensor element.

5. Conclusions

In this paper we presented a novel low-cost silicon surface-micromachined angular rate sensor. This sensor has been conceived for rollover sensing, but – after optimisation – the use in navigation control seems to be also possible. The sensor works under low pressure close to vacuum. It is sealed with a silicon micromachined cap and encapsulated in a standard PLCC44 package. The sensor shows a good linearity and allows a customer specific sensitivity of up to 7 mV/°/sec with a measurement range of +/-250 °/sec.

6. Acknowledgements

The authors are highly indebted to the FAB-team for the fabrication of the sensors. Furthermore, we would like to thank Michael Fehrenbach, Dirk Scholz, Michael Offenberg, Frieder Haag, Peter Hein, Frank Henning, Frank Fischer and Doris Schielein for their work and for helpful advices.

7. References

[1] J. Marek. Silicon microsystems for automotive applications, Robert Bosch GmbH, 1997.

[2] C. Song, M. Shinn, Commercial vision of silicon-based inertial sensors, Sensors and Actuators A66(1998), pp. 231-236

[3] J.-J. Choi, R. Risaku, K. Minami, M. Esashi, Silicon Angular Resonance Gyroscope by Deep ICPRIE and XeF2 Gas Etching, IEEE Proceedings of MEMS98, Heidelberg, pp.322-327

[4] W.A. Clark , R.T. Howe and R. Horowitz, Surface Micromachined Z-Axis Vibratory Gyroscope, IEEE

[5] B. Folkmer, J. Merz, H. Sandmaier, W. Lang, A New Silicon Rate Gyroscope, W. Geiger, IEEE Proceedings of MEMS98, Heidelberg, pp. 615-620

[6] M. Lutz, W. Golderer. J. Gerstenmeier, J. Marek, B. Maihöfer, S. Mahler; H. Münzel, U. Bischof:A Precision Yaw Rate Sensor in Silicon Surface Micromachining, Proceedings of Transducers '97

[7] M. Offenberg, F. Lärmer, B. Elsner, H. Münzel, W. Riethmüller, Novel Process for a Monolithical Integrated Accelerometer, Transducer'95, Eurosensor IX, 148-C4, pp. 589-592

[8] M. Offenberg, H. Münzel, D. Schubert, O. Schatz, F. Lärmer, E. Müller, B. Maihöfer, and J. Marek, Acceleration Sensor in Surface Micromachining for Airbag Applications with High Signal/Noise Ratio, Robert Bosch GmbH, SAE 960758,1996

[9] D. Teegarden, G.Lorenz, R. Neul. How to model and simulate microgyroscope systems, IEEE Spectrum, pp.66-75, July 1998

A New Generation of Micromachined Accelerometers For Airbag Applications

M. Aikele, M. Rose, R. Gottinger, U. Prechtel*, J. Schalk*, T. Ohgke,
M. Weinacht, and H. Seidel
Temic, Product Unit Sensor Systems, D-81663 Munich, Germany
*DaimlerChrysler, Research and Technology, D-81663 Munich, Germany

Abstract: In this contribution an overview concerning the design, fabrication and performance of a new generation of micromachined silicon accelerometers will be given.
First we focus on two- and three-axes bulk-micromachined accelerometers with piezoresistive read-out. The two-axes version represents a cost-effective solution for front and side impact detection in the automobile. The monolithically integrated triaxial sensor additionally includes the possibility for roll-over detection. It consists of four seismic masses with the main axis of sensitivity showing an inclined angle with respect to the chip normal.
In a second part we present a completely new resonant accelerometer principle. The resonator consists of a double clamped beam coupled to a seismic mass. The beam is excited thermally by an implanted resistor and sensed piezoresistively. The first resonance frequency of the beam is about 400 kHz. An acceleration leads to a tensile or compressive strain inside the resonator and shifts the resonance frequency. We have measured a sensitivity of 70 Hz/g and a quality factor of 3000 at a pressure of 1 mbar. The principle layout of the electronic circuitry will also be presented.

Keywords: Acceleration Sensor, Multi-axial Sensor, Resonant Sensor, Silicon Micromachining

1. Introduction

Microsystems technology can fully deploy its capabilities for fabricating highly reliable products at moderate costs when serving high volume markets. One of the most important is the automobile, which requires an increasing number of sensors to satisfy novel functions, concerning the areas of motor management, environmental control, as well as drivers safety and comfort. In the last several years, the airbag has become a common product in most cars, which can increase the safety of the driver and the occupants considerably in crash situations. The heart of the electronic control unit is an accelerometer. In many cases, an additional threshold switch for acceleration (safing sensor) is also included for safety reasons. In the last few years, conventional sensors are beginning to be replaced by micromachined silicon sensors for reasons of reduced cost and additional performance.

2. System Aspects

In the past, the usual configuration of an airbag deployment system consisted out of two frontal airbags and two belt pretensioners for the driver and his co-passenger. The measurement of two axes is now almost standard for the central control unit. This system served in front and rear crash situations, and typically required two accelerometers and a safing sensor at the central single point detection location. Recently, side airbags were added in order to reduce the injury in side impact crash situations [1]. Since a side impact leaves less time for the system to react, the addition of two satellite accelerometer units was required, which are positioned in the door frames. Each sensor is equipped with its own microcontroller.
The effectiveness of present airbag systems has substantially lowered the number of injuries due to front and side impact crashes. Thus, however, the relative importance of roll over crashes has increased, since no protection is provided for this event, up to now. For future systems it is planned to include a head airbag, which will require an additional low-g z-axis accelerometer and an angular rate sensor for the longitudinal axis of the car. In the following, we present an accelerometer which can measure all three axes, monolithically.

3. Multi-Axial Accelerometer

The usual approach for a two-axial sensor is the application of two individual sensor elements with their sensitivity axis within the plane of the wafer surface [3]. One is rotated by 90° with respect to the other. This concept fails, when a third axis is to be included, since a totally different type of sensor is required for the third axis. Our approach, in contrast, is the application of a sensor element with its main axis of sensitivity inclined to the wafer normal. This is the type of sensor that we use as the standard single-axis airbag accelerometer [2]. By using two or four identical seismic masses with the suspending cantilever beams rotated 180° or 90° to each other, respectively, a two-axial, as well as a fully three-axial sensor can be fabricated, monolithically. Each sensor pair serves to detect two acceleration axes, one vertical and the other parallel to the wafer plane. This concept is schematically shown in Fig. 1.

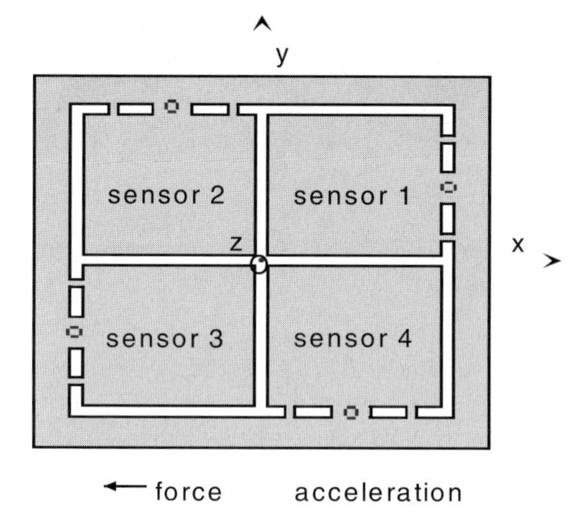

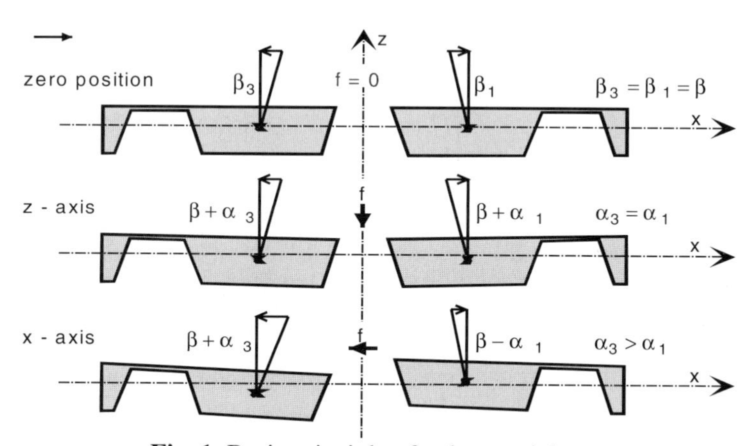

Fig. 1: Basic principle of a three-axial sensor.

An actual scanning electron micrograph of a three-axial sensor is shown in Fig. 2. Four seismic masses are positioned over one single etch cavity. From these four elements the three components of acceleration, a_x, a_y, a_z can be extracted from the sensor signals by forming sums and differences between opposite pairs, according to the following equations:

$$a_x = \frac{1}{2(k^2 + q^2)}[k(U_3 - U_1) + k(U_4 - U_2)] \tag{1}$$

$$a_y = \frac{1}{2(k^2 + q^2)}[k(U_4 - U_2) + k(U_3 - U_1)] \tag{2}$$

Fig. 2: Scanning electron micrograph of a triaxial accellerometer.

$$a_z = \frac{1}{4A}\left[U_1 + U_3\right] = \frac{1}{4A}\left[U_2 + U_4\right] \qquad (3)$$

where U_i are the four sensor output signals, A is the sensitivity in the main axis, k is the sensitivity in the longitudinal axis and q is the sensitivity in the transverse axis. The magnitude of the coefficient k is determined by the alignment angle of the main axis of sensitivity with respect to the wafer normal. For typical angles of about 20°, it reaches a value of 30-40% of the main axis of sensitivity. The transverse axis term q depends on the sensor configuration and was typically measured to be about 1% or even lower. So it can be neglected in cases where the required precision is not too high. Thus, the equations presented above simplify considerably.
From Eq. 3 it can be seen that the z-component of acceleration can be determined in two independent ways, by both sensor pairs. Mathematically, this follows from the fact that three components of acceleration are measured with four independent sensor signals. This redundancy gives the possibility to perform an ongoing self monitoring consistency check of the sensor and recognize faulty situations.

The result of a typical measurement by turning the device around its x-axis within the field of gravity of the earth is shown in Fig. 3. The inner lines which cross over at 0° and at 180° show the four signals of the individual sensors (two of them are practically identical). From these, the three components of acceleration are extracted by applying the above mentioned equations. The y- and the z-components follow the sinusoidal motion with a phase shift of 90°. The x-component is nearly zero, as would be expected.

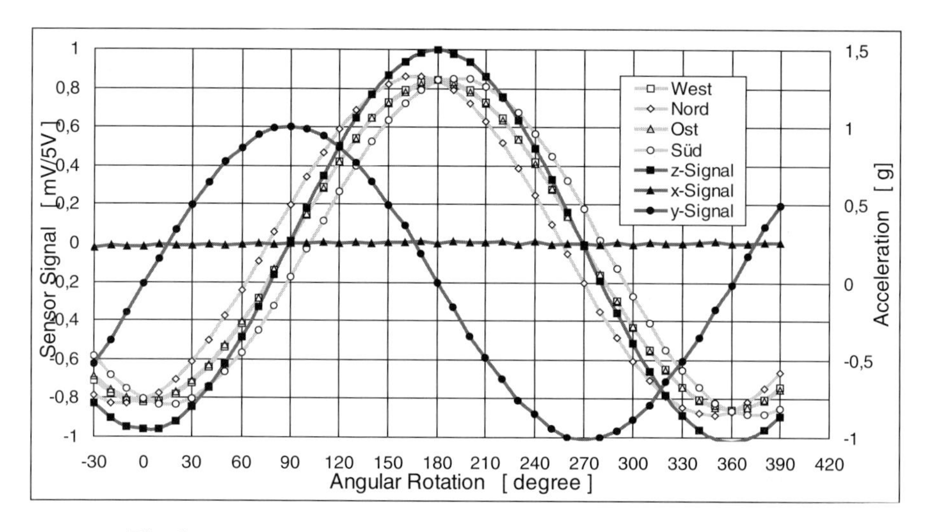

Fig. 3: Measurement results of the triaxial sensor being rotated in the field of gravity of the earth around its x-axis.

The start of the high volume production of a two-axial sensor with one pair of masses is planed for the year 2000. Prototypes are already available. It is completely the same technology as our single-axis airbag accelerometer being in production. The dimensions at the top of each mass are 560 x 1130 μm². The overall size of this sensor lies a factor 1.5 over the single-axis sensor . Therefore it is a very cost-effective two-axes accelerometer.

4. Accelerometers With Resonant Read-Out

4.1 Function Principle and Design

A relatively new approach for a micromechanical silicon accelerometer is to use a resonant signal pick-up concept [4].
The resonator consists of a double clamped beam with a length of about 200 μm and a width of approximately 3 μm. The beam is excited thermally by an implanted resistor and sensed piezoresistively. The resonance frequency of the first mode of the beam lies at approximately 400 kHz. This is in good agreement with analytical calculations and with finite element simulations. For excitation we use a sinusoidal voltage. It is clear that each halfwave of this voltage leads to an expansion of the material. This results in a frequency doubling by the thermal

Fig. 4: Scanning electron micrograph of a silicon accelerometer with resonant read-out.

excitation itself. The more mathematical argumentation says that the power is proportional to the square of the voltage leading to a frequency doubling. Alternatively the sensor can be excited with its real resonance frequency by adding an offset voltage. In this case the power consumption increases. The second mode of the resonator at approximately 1.2 MHz can be excited, too.

Fig. 4 shows a micrograph of a realized device. The seimic mass with lateral dimensions of 400 x 400 μm^2 is located in the center. This mass can move laterally and is coupled via its suspending cantilever to the double clamped resonator beam. The complete mass-spring system has a resonance frequency of about 17 kHz. This lies well above the vibratory spectrum of the automobile. In case of an acceleration, the mass moves laterally and induces a tensile or compressive strain inside the resonator which leads to a shift of its resonance frequency. The acceleration force of the seismic mass on the resonator is amplified by the leverage ratio of the cantilever. A value of approximately 10:1 can be realized.

4.2 Measurement Results

Fig. 5 shows the excitation in resonance as a function of the frequency at five different energy levels. The frequency shift towards lower values with increasing power is due to a rise of temperature in the thermal excitation structure. For practical purposes, an input power of about 1 mW is sufficient. The quality factor

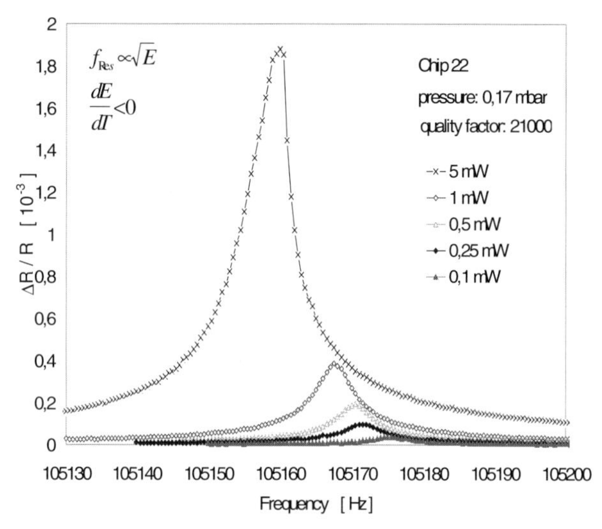

Fig. 5: Piezoresistive output signal as a function of frequency at different excitation energy levels.

Q is a strong function of the pressure within the air-gap. At pressures below 1 mbar we typically reach a value between 10000 and 30000. The highest observed value was close to 100000. The maximum pressure at which this device can be operated is about 10 mbar. This is a realistic level for production.

The result of a measurement under acceleration of ± 1g is shown in Fig. 6. The measured sensitivity lies at 70 Hz/g. For the temperature coefficient of the frequency offset we expect about 5 ppm/K. The temperature coefficient of the sensitivity is expected to be negligible.

4.3 Electronic Circuitry

The resonator must be driven in resonance. The main characteristic of the resonance drive is the 90° phase shift between excitation and detection signal. This condition is fulfiled by the phase locked loop oscillator (PLL). The comparator (CMP) supplies a signal depending on the phase difference. It controls the frequency of the voltage controlled oscillator (VCO). The damping of the PLL can be tuned by the low pass filter (LP). The most effective excitation can be achieved by a sinusoidal signal.
The output signal of the sensor is amplified and filtered by the bandpass (BP) to reduce noise. This sinusoidal signal is converted to a TTL-signal. Its frequency is divided by 2, since the thermal excitation doubles the frequency.

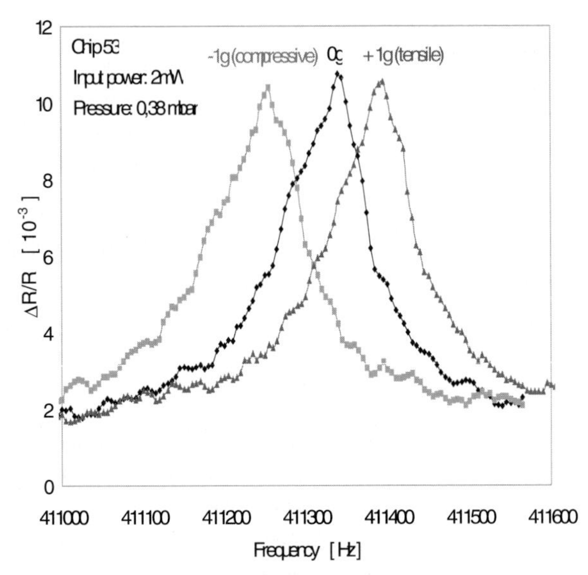

Fig. 6: Measured output signals for different acceleration levels. The sensitivity is 70 Hz/g.

On the sensor there is not only the mechanical coupling between excitation and detection. There is also a thermal crosstalking which has to be reduced. Since this crosstalking is nearly frequency independent within the working range it can be eliminated very simple by subtracting a signal of same amplitude and 180° phase shift.

4.4 Perspectives

The major advantages are that the chip size can be reduced substantially, the pick-up can be realized by a low impedance device, and the technology can be made very simple. For excitation and detection two standard implanted resistors are used. From the fabrication point of view, a quasi-surface micromachining technology in single crystal silicon can be realized. Possible thermal drifts of the piezoresistor are not relevant, since only the frequency is a measure of the acceleration.

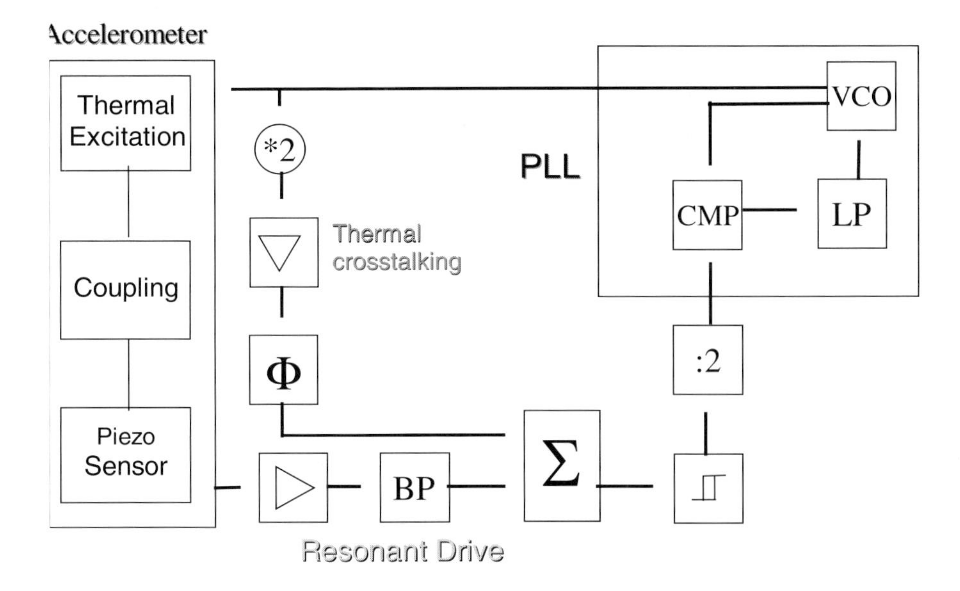

Fig. 7: Principle layout of the electronic circuitry.

Acknowledgement

This work was supported by the German ministry of research and education (BMBF) by the SiFa-project. We acknowledge the cooperation with the Fraunhofer Institute of Solid-state Technology (IFT), Munich.

References

[1] P. Steiner and S. M. Schwehr, „Future applications of microsystem technologies in automotive safety systems“, in Advanced Microsystems for Automotive Applications 98, D. E. Ricken and W. Gessner, edts., Springer, Berlin 1998, pp. 21-42.

[2] H. Seidel, U. Fritsch, R. Gottinger, and J. Schalk, „A piezoresistive silicon acclerometer with monolithically integrated CMOS-circuitry“, in Proceedings Transducers ´95, pp. 597-600, Stockholm, June 25-29, 1995.

[3] C. Lemaire and B. Sulouff, „Surface Micromachined Sensors for Vehicle Navigation Systems“, in Advanced Microsystems for Automotive Applications 98, D. E. Ricken and W. Gessner, edts., Springer, Berlin 1998, pp. 103-112.

[4] T. A. Roessig, R. T. Howe, A. P. Pisano, and J. H. Smith, „Surface-Micromachined Resonant Accelerometer“, in Proceedings Transducers ′97, pp. 859-862, Chicago, IL,1997.

Integrated Surface Micromachined Gyro and Accelerometers for Automotive Sensor Applications

Bob Sulouff, Analog Devices Inc, 21 Osborne St., Cambridge, MA

Abstract. The application of integrated surface micromachining for improved automotive gyros and accelerometers is discussed and evaluated. The value of single chip or integrated micromachined sensors has been questioned by sensor development technologist's due to perceived complexity and high development costs. This paper uses the recently developed single chip micromachined gyro and single chip accelerometers to evaluate the trade-off in integrated structures. The product design and manufacturability of the gyro is reviewed in detail. The scaling to low signal levels (20zepto farads, 10^{-21}) with a total full scale change of 10 femto-farads is only possible when the electronics and associated parasitics are integrated such as the gyro. The trade-off of integrated micromachined sensors for cost and reliability are also discussed. The paper concludes with the applications of the gyro and associated low noise and higher performance accelerometers for vehicle dynamic control, roll-over and navigation applications. These applications insure a new capability in intelligent power train and safety applications.

Keywords: Gyroscope, Surface Micromachining, Navigation, VDC

1. Introduction

The integration of mechanical and electronic functions has been a benefit to automotive system designers for many years. The sensor field is an area where recent advances in the technology have created opportunity and at the same time challenges for integration. The introduction of surface micromachining has made the potential for integration on a single monolithic substrate a reality with some unique benefits that are only possible when the mechanical structure is combined with the electronics. The inertial sensing field has a high demand for innovative cost-effective solutions and does not require media compatibility such as a pressure sensor or flow sensor. By using standard IC processing techniques, surface micromachining can follow the same economic scaling trend that modern day integrated circuits has taken[1,2,3]. Having made the investment in developing an integrated process, additional electronics or levels of mechanical complexity are small incremental costs that provide improved multiples of sensing direction or function. An added benefit of single chip sensors is the improvement in device consistency as the structures share the same thermal and stress environment. The tighter matching of mechanical features, that occurs when structures are manufactured using the same process and tools results in less variation. The automotive applications also require price, performance, and reliability levels that are difficult to achieve when multiple assembly operations must be handled and yielded separately with a multi-chip approach. A scaling run-away occurs when larger sensors are created to provide additional signal to overcome interconnections and parasitics. The electronics which are now on a different chip must become larger to add the additional EPROM circuitry to compensate for stress and temperature variations. The interaction of different signals both within the sensor and fields from outside the sensor are a greater concern when long interconnections are used. A single small chip in a conventional package provides a cost and reliability solution for inertial sensors such as gyros and accelerometers.

2. Surface Micromachined Gyro

The integration of a complete gyro function on a single chip has been accomplished by Analog Devices [4]. The gyro has several unique features that can be noted in figure 1. All of the electronics for the sensor are included on the 3 mm by 4 mm silicon chip. The design uses two accelerometers that are driven by a comb structure to create a reference velocity. This structure is built from a 2 micron thick polysilicon surface micromachined process that is integrated with BiMOS circuitry. Thin film SiCr resistors are included and trimmed at the wafer level to give a compensated analog output over temperature. The sensing direction of the device is about the vertical axis (yaw). The accuracy is 0.5 deg per second with a full-scale range of ± 300 deg per second. To provide additional

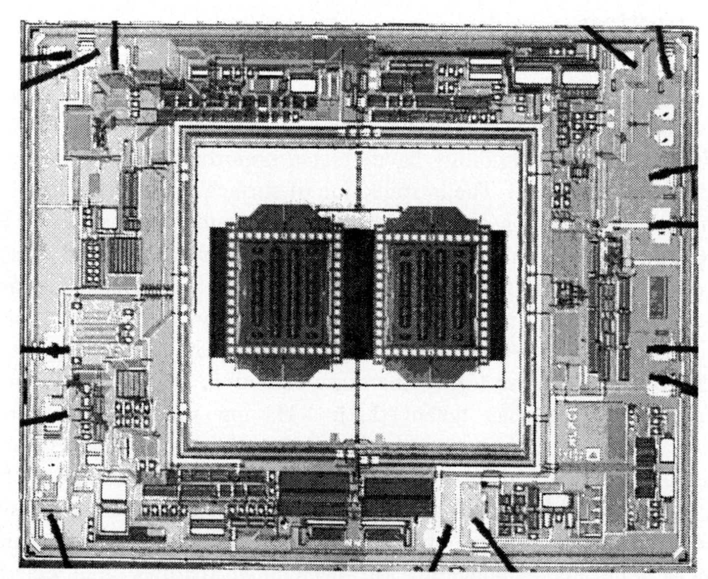

Fig. 1: Integrated Surface Micromachined Gyro (Yaw Rate)

information the gyro has a temperature sensor which gives the silicon die temperature as a voltage output. An important feature of the integrated design is the level of mechanical optimization possible. Many drive fingers (5000) provide force to move the sensing accelerometers even in a dry air filled package. With the design of a mechanical coupling mechanism the rejection of quadrature interference has been demonstrated to be 0.7 ppm. This makes the demodulator phase accuracy easy to achieve. Micromachined gyros must detect a small signal in the presence of the vibratory velocity. A vibration at 16 kHz and a displacement of 10 microns results in an acceleration of 100 km/s^2 (10.2 kgees) while the Coriolis acceleration that results from 1 deg per second would be 35 mm/s^2 or about 4 milligee. The challenge is to detect this signal which is 1/3 of a part per million of the excitation. Figure 2a shows the output of the gyro in volts as the rate is changed. The error of the output signal verses the rate is shown in figure 2b. The major contributor to the nonlinearity is the demodulator electronics which has been improved since this data was taken. With the quadrature interference at 0.7 ppm, improved electronics could be used for higher accuracy.

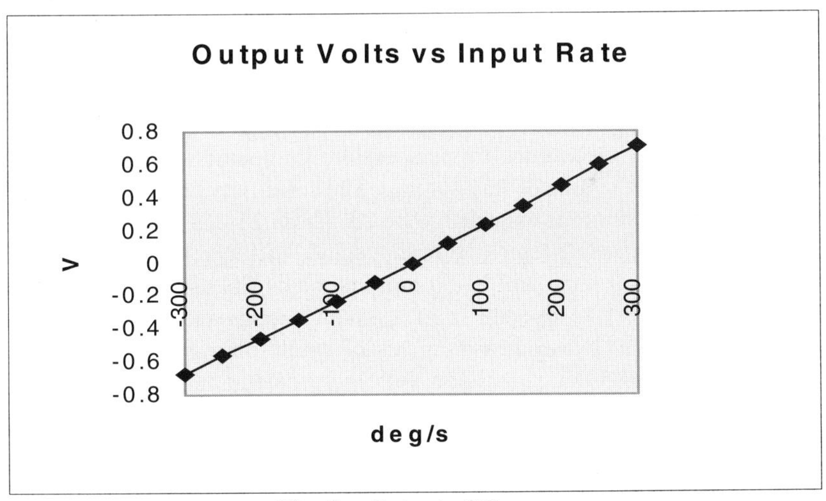

Fig. 2a: Output of Gyro

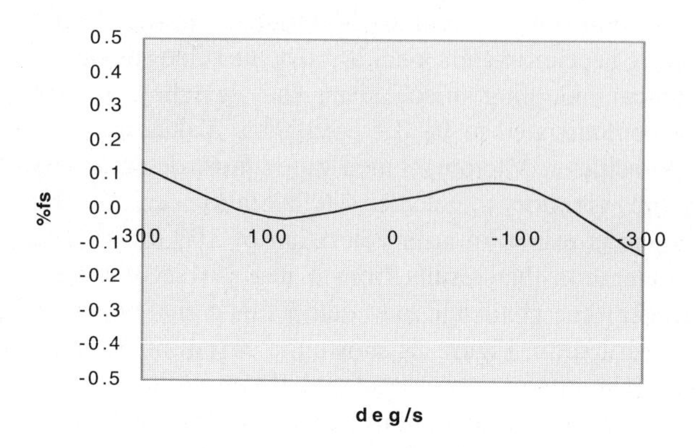

Fig. 2b: Error of Output

The advantages of the single chip integrated design are evident in the measured results of the electronics feed-through. Only 0.3 ppm of feed-through occurred which is 100 times less than a two chip design with the typical lengths of interconnection between the sensor and the electronics. The null stability for a test time of 1 hour was measured to be less than 0.1 deg per second at 25 deg C. The optimization of the electronics for temperature compensation and start-up time were in progress at the time of this writing. The sensitivity change from –40 to 85 deg C due to temperature is expected to be 2% (from 25 deg C) or less, while the null change over the same temperature range and reference is expected to be less than 3%. The turn-on time will be 50 milliseconds. The use of BiMOS circuitry has shown the ability to operate from a 5 volt supply and require less than 5 milliamps of current. Noise measurements of single chip gyros have confirmed expectations of < 0.05 deg per rt. sec. with the acceleration sensitivity below the noise for a $\pm$ 1 gee measurement. Self test or diagnostic functions are included to give a higher confidence of proper operation.

3. Manufacturability and Reliability

The use of an existing manufacturing process that produces accelerometers for automotive safety applications has been the basis of gyro design at Analog Devices since the initial concept models. The careful control of mechanical features to insure that acceleration from the vibrating function does not result in signals orthogonal to the direction of vibratory motion has been taken into considered [5,6,7]. Process characterization of polysilicon curvature is an important aspect of both gyro's and accelerometers. With excessive curvature the out of plane motion becomes significant. This has been a consideration for manufacturability [8]. Since the method of accelerometer trimming is with lasers at the wafer level, an excellent history of polysilicon properties as related to manufacturing parameters has been studied and characterized. Wafer level trimming provides valuable intermediate information on accelerometers and gyros which are further influenced by packaging stresses and thermal treatments. Unlike EPROM post package trimming, this common alternative to error correction gives the opportunity to follow trends per wafer, die site or intermediate manufacturing characteristics and unlike EPROM does not need to deal with the uncertainty produced by the packaging hystersis and creep. Analog Devices has continued to improve the polysilicon deposition and wafer level trimming technology to meet low-g accelerometer designs which are more demanding than crash sensor (50 gee). The gyro has benefited from 5 years of accelerometer production and manufacturability optimization.

Being a single chip that is die attached into a very conventional hermetic ceramic package that does not require vacuum has contributed greatly to both manufacturability and reliability. Because the gyro and accelerometers are single chip the interconnections between high impedance nodes are extremely reliable and not subject to interference or changing parasitics. As is common in integrated circuit technology the interconnections are passivated and protected. In the complete history of single chip accelerometer qualification and reliability testing at Analog Devices no failures have occurred due to silicon wafer level interconnections or their IC related processing. The common reliability issues have repeatedly been packaging and wire bonding related. This supports the value of a single chip where less devices in a package and fewer chip to package interconnections improve the reliability.

4. Integration Trade-offs

There are many factors that must be considered when a single chip verses a two-chip approach are being evaluated [9]. The general categories are:

4.1 Design

The design is defined by the manufacturing process and can vary significantly with the process capabilities. Single chip designs depend upon consistency due to tracking and the ability to measure small values in relation to non-changing references.

Accelerometer masses that are less than 1 microgram are common in single chip designs. With the ability to manage high impedance nodes and parasitics, very small capacitance values can be measured. The XL202 reported on at a previous

AMAA [10] conference has a detectable capacitance value of 20 zepto(10^{-21}) farads resulting in resolution to 2.5 picometer movement.

In contrast to the very small masses and signals that can be managed with a single chip device, two chip designs must account for the unknowns of parasitic signals created during the transfer of very low signals with high impedance nodes and susceptibility to stray signals. This drives the two-chip design to increase the sensor size and therefore the signal. Capacitive inertial devices for two chip designs must also manage the packaging of sensitive mechanical structures. The cover or capping of the sensor element can also add significantly to the device size due the bonding area and therefore increasing the opportunity for signal leakage, stresses , thermal mismatch and loss of hermeticity [11].

4.2 Manufacturability

Tools developed for integrated circuits have been applied directly to the manufacture of single chip integrated surface micromachined sensors. The etching of trenches for capacitive elements of IC's and the selective plasma etching of polysilicon has made the addition of mechanical structures logical extensions of equipment and processes already in production. The silicon on insulator technology that has made new types of devices possible has also shown promise as a substrate for sensors. With the insulator below the surface of a single crystal silicon layer, structures can be fashioned with both mechanical and electronic properties. The use of deep reactive ion etching has made the larger structures needed for the two chip approach more practical. This has helped to justify the larger devices that have larger signals. The difficulty is that with the larger structures comes additional stress and variations [12,13]. The variations are the most significant factors for manufacturability. Analog Devices has seen the advantage of small structures that are spaced closely together. As the structures get larger the variations increase. Single chip designs encourage symmetry, small size and the "cross-quading" as commonly done in differential input pairs of IC's for better matching and manufacturability. This also allows for better matching of sidewall slopes, widths and spaces and curvature management.

4.3 Cost

The pricing expectations in automotive and consumer inertial products has been toward lower prices with increased functionality. The trends in integrated circuitry that have provided for mixed signal devices with increasing performance and reduced price can be applied to integrated sensors. A major driver of costs is yield. The common argument for multiple devices over single chip devices is the opportunity to test functions separately and therefore yield the final assembly higher. The yield models when evaluated with mature products that are photo defect limited readily show the advantage of an IC process that is robust for its individual steps and can be cost effectively sequenced into a manufacturing process. The experience at Analog Devices from over 30 years of mixed signal IC production and 5 years of integrated sensors has supported the single chip solutions first in converters and now for accelerometers.

The most important factor once the yields are stabilized is the size of the die or the gross die per wafer. The major cost of a sensor is the silicon cost when the packaging and testing are high volume and low cost orientated. The initial single chip accelerometer (XL50) introduced in 1993 [9] was replaced with a die of half the size two years later and the trend is continuing with multiple axis per part and design concepts that support further reductions in die size to even 25% of the original XL50's size. When two chip devices are compared for efficient use of silicon, factors of 5 to 10 larger areas are noted which can limit the materials cost reduction of a sensor. The two chip plus capping designs typically require more masking levels, have larger silicon area by factors of 5 or 10 and require bonding and interconnection on several levels. These costs as well as the losses due to larger die that are more sensitive to the packaging stress result in costs that encourage an integrated single chip solution.

Development costs and timing are also an important part of the trade-off in two chip verses single chip. Since in both cases silicon design, photo tools and fab cycle times are involved the development costs and timing are not significantly different. The use of lower mask count mixed signal IC's that are then interconnected to the sensor cells is a more predictable and lower cost development path for the electronics. However since time to market is a critical factor as well as engineering tooling the unknowns of the sensor cell and the interaction of larger and multiple die with the packaging stresses add additional challenges to the development. The single chip approach can apply a wide variety of design simulation and modeling tools common in IC processing. The use of wafers that are used as experiments are also an effective method used at Analog Devices. On a single wafer mask set up to 20 different sensor designs have been evaluated. These multi-project runs many of which have been funded by DARPA have made it possible to develop new single chip sensor designs at low cost and with rapid time to market.

4.4 Quality and Reliability

The trade-offs of single verses two chip sensors in the area of quality and reliability are difficult to perform. A single chip with it's reliable on chip interconnections and passivations simplify the product and give somewhat of an advantage. The two chip versions have like wise demonstrated good quality and reliability. Since without acceptable single digit ppm's and long term reliability a sensor would not be qualified for automotive, the trade-off is in the cost to manufacture a high quality product. This discussion is therefore more a consideration in the cost and manufacturability areas.

4.5 System Requirements and Growth

As the smart air bag becomes a common product, the system requirements have placed additional demands on the sensor. The single axis sensor that meets system needs a few years ago is now only part of an expanded system. The side air bag and associated frontal crush zone sensor as well as new methods to provide safing functions with electronics has created needs for multiple axis devices. Here the single chip solution has demonstrated some significant advantages. The use of photolithography can keep alignment between multiple sensors on the some chip well within a fraction of a degree. The use of common electronics also provides good leverage for multiple axis parts. The gyro discussed in an earlier section of this paper is applicable to rollover detection. The design of multiple gyro axes and acceleration sensors have already been show to be possible on a single chip. The challenge is to apply manufacturability and cost considerations to meet the necessary system requirements. As the complexity of the functions increase to meet expanded system requirements the economic solution for sensors appears to parallel the modern day microprocessor. More integration and a more universal product develops using the technology of integrated circuit processing.

5. Applications

The automotive applications of integrated surface micromachined gyro's and accelerometers have been demonstrated in air bag systems. The cost, performance, quality and reliability have been tested and proved in very large quantities over the last 6 years with 10's of millions of accelerometers meeting very high expectations. The applications in the safety system area are continuing to increase. The side impact requirements have increased the need for two axis devices. The addition of the variable inflation rate bags has increased the need for the frontal crush zone sensor. The use of multiple sensors for electronic sensing has also been part of opportunities for integrated surface micromachined sensors. With the rollover system [14] the gyro and low-g accelerometers have become very important sensors.

The use of gyro's as yaw rate sensors for vehicle dynamic control has been demonstrated in production in the last few years. The trend is to an ever-increasing need to determine inertial forces on the vehicle and improve the handling safety and performance. Low-g accelerometers for lateral and longitudinal acceleration sensing are also required for vehicle dynamics.

The use of GPS for car navigation and the application of gyro's and accelerometers to supplement the GPS signal are currently in use. The larger piezoelectric based gyro's are expected to be replaced with smaller gyro's that can be mounted into the GPS receivers.

6. Acknowledgements

The author would like to express his sincere thanks to John Geen who is the designer and lead technical representative for ADI's gyro activity. John Chang has also contributed significantly to the gyro efforts along with Steve Lewis and Steve Sherman. Bill Riedel's helpful discussions and review of the manuscript are appreciated. The author would also like to acknowledge the US Government for DARPA/TRP funding.

7. References

[1] R.T. Howe, Polysilicon integrated microsystems: technologies and applications, in: Proceedings of Transducers '95, Stockholm, 1995, pp.43-46.

[2] W.Yum, R.T. Howe, P.R. Gray, Surface micromachined, digitally force-balanced accelerometer with integrated CMOS detection circuitry, in IEEE Proc. of Solid State Sensors and Accuators Workshop, Hilton Head , 1992, pp. 126-131.

[3] M.A. Lemkin, B.E.Boser, D. Auslander, J.H.Smith, A 3-axis force balanced accelerometer using a single proof-mass, in: Proceedings of Transducers '97, Chicago, 1977, pp. 1185-1188.

[4] J.Geen, A path to low cost gyroscopy,in: Solid State Sensors and Actuators Workshop, Hilton Head , 1998, pp. 51-54

[5] M.Putty, K.Najafi, A micromachined vibrating ring gyroscope, in: IEEE Proceedings of the Solid State Sensors and Actuators Workshop, Hilton Head, 1994, pp 213-230

[6] M. Lutz, W. Golderer, J.Gerstenmeire, J. Marek, B.Maihifer, S. Mahler, H. Munzel, U.Bischof, A precision yaw rate sensor in silicon micromachining, in: Proceedings of Transducers '97, Chicago, 1977, pp. 847-850

[7] C. Song, Commercial vision of silicon based inertial sensors, in: Proceedings of Transducers '97, Chicago, 1977, pp. 839-842.

[8] K.Chau, R. Sulouff Jr, Technology for the high-volume manufacturing of integrated surface-micromachined accelerometer products, in: Microelectronics Journal, vol 29, (1998), pp.579-586

[9] R.Payne, S.Sherman,S.Lewis,R.T.Howe, Surface micromachining: from vision to reality to vision, in: IEEE Proce.of the ISSCC'95, San Francisco, 1995,pp. 164-165.

[10] C.Lemaire, R. Sulouff, Surface micromachined sensors for vehicle and personal navigation systems, AMAA Proceedings, 1998, Berlin, pg 23-26

[11] W.Golderer, M.Lutz, J.Gerstenmeier, J.Marek, Maihofer, S. Mahler, H. Munzel, U. Bischof, Yaw rate sensor in silicon micromachining technology for automotive applications, AMAA Procedings, 1998, Berlin, pg. 69-78

[12] Motorola Data Sheet, MMAS40G10D, Micromachined Accelerometer, 1997

[13] I.Ristic, D. Koury, E. Joseph, F. Shemansky, M. Kniffin, A two-chip accelerometer system for automotive applications, Micro Systems Technology '94, Berlin, 1994, pp. 77-81

[14] G.Mehler, B.Mattes, M.Henne, H-P Lang, W. Wottreng, Rollover Sensing (ROSE), AMAA Procedings, 1998, Berlin, pg 55-68

A New Microelements Electrostatic Actuator for Automotive Applications

Marco PIZZI, Elena BASSINO , Sabino SINESI, Pietro PERLO, Valerian KONIACHKINE*
FIAT RESEARCH CENTER
OPTOMECHANICAL TECHNOLOGIES
Strada Torino 50, 10043 Orbassano (TO), Italy.
*Siberian Branch of Russian Academy of Sciences, Sobolev Institute, acad. Koptuga 4, 63009 Novosibirsk, Russia.

Abstract: In this paper we describe a new type of electrostatic actuator, based on the parallel action of many micro elements, that allows high versatility, adjustable speed (included very low if required), high precision displacement, low weight and very flat design. The high level of integration of the proposed solution and the criteria on the choice of the working principle are reported as well.

Keywords: Electrostatic Actuator, PET, Linear Motion, Rotative Motion

Background

In a medium class car, there are tens of electric motors in a range of power from a fraction of watt to few hundred watts. The most widely used working principle of these motors is the electromagnetic induction. The general criteria followed to develop new forms of actuators are:

- functional improvement
- cost reduction
- weight reduction
- volume reduction .

As the motors operate independently and only periodically, the current trend it is to consider their efficiency only of secondary importance, but this situation may change in association with the number of motors in use at the same time.

The first step is the choice of the motor based on the best working principle as well as its level of technological integration with the rest of the system.
In Fig. 1 we show the experimental dependency of power density versus power for some kind of actuators with different working principles. A similar dependency have been presented by Dyatlov, et al. [1]. This graph is even more

relevant if considered as an equivalent representation of the relation between efficiency and power.

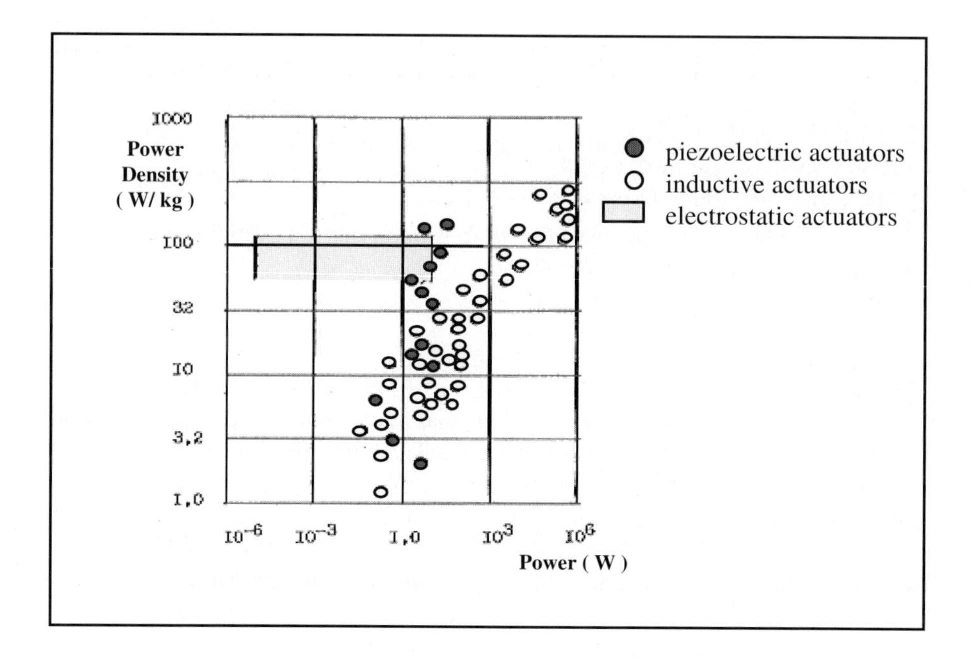

Fig.1 Dependency of power/mass from power for different kind of actuators

We can see that essentially all form of electromagnetic actuators have a lower power density at low power: the reason of this is rather physical than technological.
In particular if we suppose to use the same materials to manufacture motors of any dimension the theoretical progress of power density is [2] :

$$\frac{\langle W \rangle}{M} \propto \langle W \rangle^{\frac{1}{4}}$$

and if we consider only energy losses due to the Joule effect (negligible for superconductors) the relationship between the efficiency and the power is analogous to the previous:

$$\eta \propto \langle W \rangle^{\frac{1}{4}}$$

Regarding weight and volume reduction we look for the possible highest power densities expressed as power/mass.
At higher power level the efficiency of the inductive motor could be higher than 90%. At these power levels it would be useless to search for better solutions based on classical materials. In fact, the current trend is toward inductive motors based on superconductors [3] which could find applications on all electric vehicles.
In the range from 1 to 10 watts piezoelectric and inductive motors have similar power densities (or equivalently for the efficiencies) depending very much on their design. The electrostatic motors, at least for the known solutions, seem to be less promising.
At power levels well below one watt the electrostatic approach is much more efficient and for a high level of miniaturisation and high power densities, the best approach for actuators seems based on simultaneous use of piezoelectric and electrostatic devices [2].
For medium and higher power applications new forms of actuators could be competitive with inductive motors only if they give a strong functional improvement in terms of flexibility and actuation simplicity [4,5].

Description of the multielements electrostatic actuator

The scaling of electrostatic force is a strong motivation for the investigation of electrostatically driven actuators on the microscale and for low power applications. Operation on a macroscopic level is obtained by networking numerous micro-scale modules so that force and displacement could be magnified.
Classical electrostatic actuators based on silicon technology are currently characterised by high speed rotation, low force and high operation voltage [6] [7]. The actuator we describe in this paper is based on the action of electrostatic elements, rolling independently but synergistically on the rotor. Fig. 2 shows the structure and the working principle of a single electrostatic element.

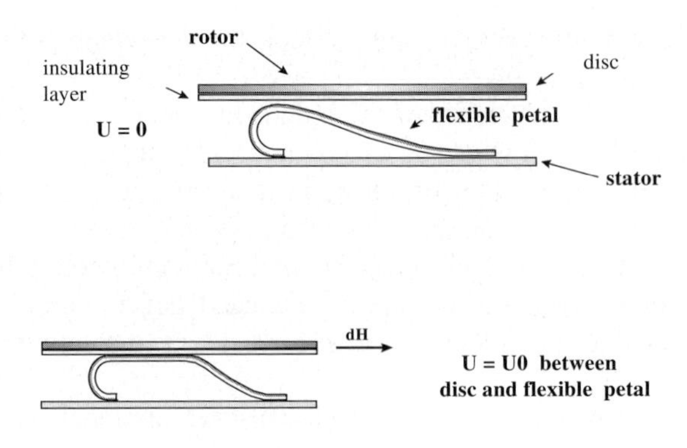

Fig. 2 : working principle for linear and rotative high resolution film electrostatic actuators

The electrostatic actuator is composed by a stator and a rotor. The rotor is a metal disc covered by an insulating layer.
The main part of the stator is the "petal" : either a metallic film fixed at its ends on a rigid substrate or a plastic flexible films (coated by a conducting layer) have been used.
Applying a voltage between the metal disc of the rotor and the petal the electrostatic adhesion occurs: the asymmetric shape of the petal gives a different mechanical tension on the left and the right end of the petal itself, forcing the rotor to move. When the system attains the desired displacement the voltage is switched off and the petal, coming into the original position, is ready for the next step.
In classical electrostatic actuators the distance between electrodes is typically quite high and this implies limited forces and high operating voltage. The motor described above represents a more efficient and powerful way to use the electrostatic force because moving electrodes are in contact with the rotor and the distance between electrodes is limited at few microns. By the choice of an appropriate dielectric it is possible to achieve high electric fields in the working region between "petals" and rotor.
To show the effectiveness of the new form of electrostatic actuator we report the behaviour of a rotary motor operating in its simplest possible construction (Fig.3). The motor whose overall thickness is 2 mm, consists on a single PET (polyethylentereftalate) aluminium coated flexible petal 6 microns thick and 8mm long.

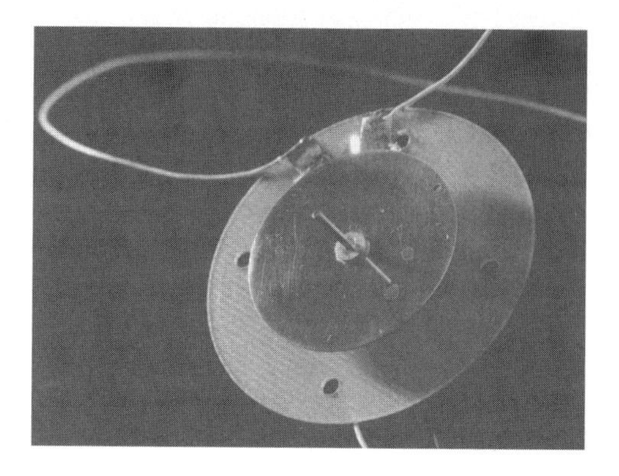

Fig. 3: Rotary motor: overall thickness 2 mm , singlepetal 6 microns thick and 8mm long; radius of motor 15 mm

The intrinsic linear motion of the petal, acting perpendicularly to the radius of the rotor , is converted in rotary motion. The accuracy of positioning depends on the size and position of the petal on the stator besides the linear displacement of the petal in a single step is dependent on frequency, as shown in Fig.4. For the specific design taken as a reference that implies that the frequency determines the number of steps necessary for single revolution of the rotor (Fig.5). The applied voltage is 200 V AC.

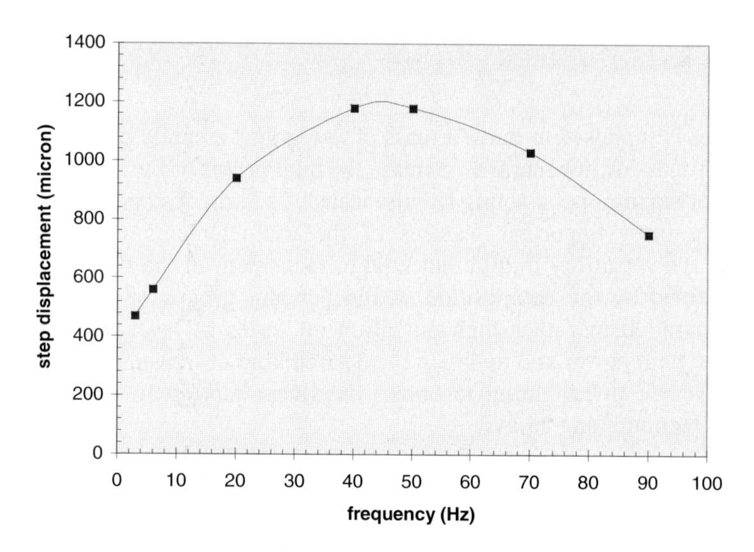

Fig.4: Single step linear displacement of the petal vs frequency

In the quasi static working region (less than 4 Hz) the minimum angular displacement is very small and specifically at 2Hz the angular displacement per step is (31.4 ± 0.4) mrad .

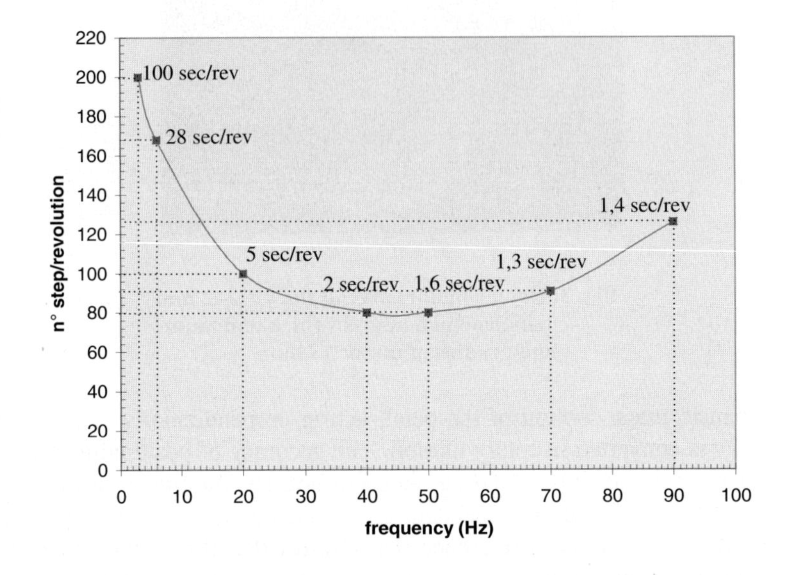

Fig.5 Number of steps for a single rotor revolution vs frequency

If the motor is supposed to move a mass of few grams a single petal is sufficient. For heavier mass or in general to increase the total power of the actuator we use a multitude of smaller petal acting simultaneously at higher frequencies.
One of the most important features of the multielements electrostatic film actuator is the versatility in different kind of movement in the three dimensional space, obtained by the composition of independent groups of petals: solutions based on parallelism rather than on piling up series of layers can be used to increase the total power and to obtain the desired kind of movement. Fig.6 shows a design we are investigating to obtain the linear motion in the positive and negative direction along an axis.

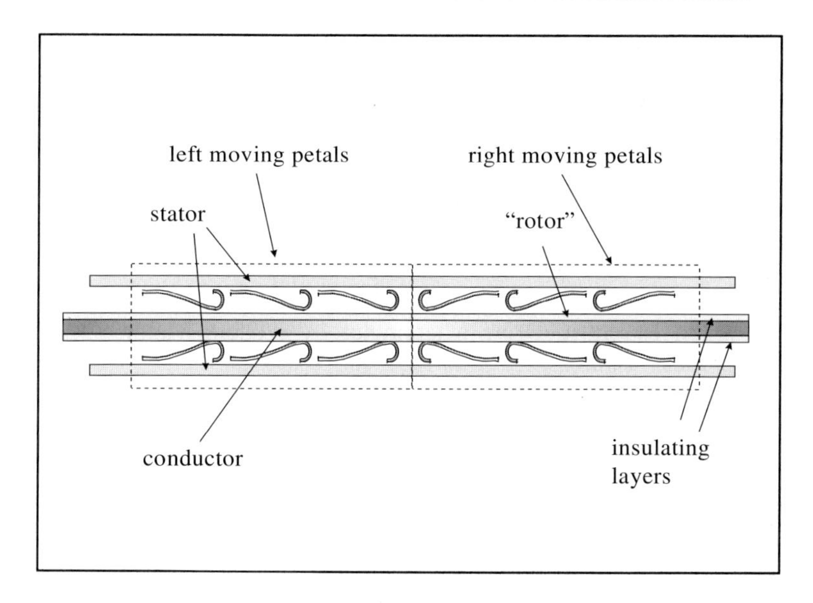

Fig.6: multielements electrostatic film actuator to obtain the linear motion in the positive and negative direction along an axis

The functional versatility and the technological integrability of the proposed actuator in the electro-mechanical systems will give a functional improvement and a cost reduction of the systems themselves.

Conclusions

The actuator described in this paper is a new way to use into the macro world the known advantages of the systems operating at micro level, in particular scaling effects, versatility , and precision displacements.
The general motivations for the development of new forms of actuators as cost reduction, functional improvement, weight and volume reduction are the main features of the innovative electrostatic actuator presented in this paper.

Reference

[1] V.L.DYATLOV, V.V.KONYASHKIN, B.S.POTAPOV, YU.A.PYANKOV. Prospects of the employment of synchrotron radiation in film electrostatic actuator technology. *Nuclear instruments and methods in physics research A 359(1995) pp.394-395.*

[2] M.PIZZI Nuove forme di attuatori a film: teoria e sviluppo di prototipi. *Tesi di Laurea in Fisica, Università degli Studi di Torino, 1996.*

[3] N.SCHIFF, R.SCHIFERL Latest Developments in Superconducting Motors. *Power transmission design, August 1995.*

[4] U.SCHAAF, G. DIEFENBACH, R. COLSON. Piezoelectric Motors for Automotive Applications. *Philips Components. Feb. 1996.*

[5] H.FUJITA. Recent Progress of Microactuators and Micromotors, *Ed H. Reichl, A. Heuberger October 1994.*

[6] R.H. PRICE, S.J.CUNNINGHAM, S.C. JACOBSEN Field Analysis for the Electrostatic Eccentric Drive Micromotor ("Wobble Motor") *Journal of Electrostatics, Vol.28, No.1, May 1992, pp. 7-38 .*

[7] H. FUJITA, A. OMODAKA Electrostatic Actuators for Micromechatronics *Proc. of the IEEE Micro Robots and Teleoperators Workshop, Hyannis, MA, 9-11 Nov. 1987*

The Application and Evaluation of a Novel Engine Management System Based on Intelligent Control and Diagnostics Algorithms and Utilising Innovative Sensor Technology

A. Truscott[1], A. Noble[1], G. Krötz[2], M. Eickhoff[2], M. Hart[2], R. Müller[2], C. Cavalloni[3]

[1]Ricardo Consulting Engineers Ltd, Bridge Works, Shoreham-by-Sea, BN43 5FG, UK
[2]DaimlerChrysler AG, T721, 70546 Stuttgart, Germany
[3]Kistler Instrumente AG, Eulachstrasse 22, PO Box 304, CH-8408 Winterthur, Switzerland

Abstract

Conventional Engine Management Systems (EMS) are primarily parameter based systems in which there is no underlying model to describe the behaviour of the engine. Such systems therefore require many parameters to control the engine under steady-state, transient and varying ambient conditions. The introduction of Model Based Control in order to overcome this problem has been hindered mainly by the lack of inexpensive sensing devices which can closely monitor the engine's performance. In particular, the measurement of cylinder pressure which would give valuable information on the engine's performance, has been very expensive due to the harsh environment of the combustion chamber. In recent years, attention has focused upon the development of inexpensive cylinder pressure sensors in order to realise a system more appropriate for the application of Model Based control. This paper describes the application of new robust pressure sensing technology coupled with Model Based control and diagnostics algorithms.

1 Introduction

Road transport world-wide is having a significant impact on global pollution levels. This is set to increase with the growth in traffic. Traditionally, any improvements in engine performance in terms of lower emissions and fuel efficiency have meant major changes in mechanical configuration. Examples of these are new types of fuelling systems, direct injection combustion systems, piston shapes, Exhaust Gas Recirculation (EGR), catalytic converters, etc. Although these have lead to noticeable performance benefits, such changes are nevertheless very costly.

In contrast, changes in engine electronics are relatively low cost. This becomes more so when a new sensor can replace a number of conventional sensors, or new software containing more advanced control algorithms is implemented. High benefit/cost ratio leads to more widespread diffusion of such technology to the market.

In recent years there has been much interest shown in the development of advanced Engine Management Systems. However, although there has been major improvements in the electronics of such systems, the advances in sensor technology and intelligent control and diagnostics algorithms has been relatively slow.

This paper describes a project underway which considers the Application and Evaluation of a Novel Engine Management System Based on Intelligent Control and Diagnostics Algorithms and Utilising Innovative Sensor Technology (AENEAS). A successful application of intelligent control, particularly model based control, requires accurate information of the combustion processes within the cylinders. Since cylinder pressures would provide a direct indication of engine load, such measurements would provide an accurate means of monitoring engine conditions. Traditionally, such devices have been very expensive to produce and durability has been a concern due to the harsh environment of the combustion chamber.

Advanced R&D work has been carried out by DaimlerChrysler and Kistler on the application of Silicon Carbide on Insulator (SiCOI) and Silicon On Insulator (SOI) technologies for in-cylinder pressure sensing devices. These material systems have been shown to operate under harsh combustion environments and have the potential, together with advanced intelligent control and diagnostics algorithms, for a cost effective solution to the development of a Cylinder Pressure based Engine Management System (CPEMS).

The AENEAS Project aims to assimilate CPEMS technology in order to obtain substantial improvements in cost, efficiency, comfort, emissions and reliability compared to conventional systems. This paper describes the background to present in-cylinder sensing technology and the approach taken by the project partners to realise benefits of CPEMS technology.

2 Background

As described in Section 1, there has been a significant improvement of engine electronics over recent years. However, the control and diagnostics algorithms which are programmed into the Electronic Control Units (ECUs) are rapidly approaching a limit. This is because conventional EMSs are primarily parameter based systems in which there is no underlying model to describe the behaviour of the engine. Such systems therefore require many parameters to control the engine under steady-state, transient and varying ambient conditions. This results in

greater calibration effort prior to production. Current technology on parameter based systems is rapidly approaching its limit where increasing effort is required to make incremental improvements for meeting legislation requirements.
The introduction of intelligent control, where the algorithm has some embedded knowledge of the engine, is required to make the necessary step change in technology. Examples of such algorithms in the context of model based control are Adaptive/Extended Kalman Filters and Neural Networks. These would provide greater scope for meeting design trade-offs with a reduced parameter set. Systems incorporating model based control require sensors to closely monitor the engine's performance. These are most effective when located in the combustion chamber.
The following subsections describe the various sensor options available based on current technology.

2.1 Ion Current Sensor Based Techniques

In recent years there has been interest in the use of Ion Current sensing technology [1, 2]. This relies on the principle of sensing the spark discharge current of a spark ignition engine in order to provide information of the combustion process. This process therefore provides an inexpensive solution to obtaining some indication of the combustion pressure. However, the main drawback associated with such a system is that it involves very complex interpretations of measured signals in order to obtain the key variables required for engine control, such as cylinder pressure. Furthermore, these signals are heavily dependent on the chemical properties of the fuel and inducted air. Another limitation of ion current sensing is that the current signal describes only the behaviour of the ions near the spark plug, not the whole cylinder.

2.2 Optical Sensor Based Luminosity Techniques

Optical sensing devices have also been considered for advanced EMS technology. However, the optical material is sensitive to rapid in-service degradation due to combustion chamber deposits, scoring and thermal cycle damage. Furthermore, as with ion current sensing, the signals obtained require complex interpretations in order to obtain the key variables required for engine control.

2.3 Pressure Sensor Based Techniques

As mentioned in the introduction, the measurement of cylinder pressures provide a direct indication of engine load. In view of this, piezoelectric and piezoresistive based pressure sensor techniques have been considered for some time in the application of control and diagnostics [3, 4]. Amongst the piezoelectric types the

ceramic sensors offer poor performance at slowly varying pressures resulting in biased measurements. Piezoelectric crystal sensors perform better than ceramic sensors (see examples in Table 2.1). However, they are too expensive for serial production.
There are a variety of piezoresistive pressure sensors which have been developed. Table 2.2 lists some examples from several companies. More details on robustness properties are indicated in Table 2.3 along with sensor configuration and chip technology (also see Figure 2.1). It becomes evident that SOI and SiCOI technology will provide greater resistance to the combustion temperature and hence greater reliability. Depending on the packaging, these sensors can be produced at practical prices.

Table 2.1 : Piezoelectric Sensors

Company	Max. Operating Temperature	Relative Price	Performance	Availability
Kistler	350°C - 400°C	high	good	available
AVL	350°C - 400°C	high	good	available

Table 2.2 : Piezoresistive Sensors

Company	Max. Operating Temp.	Relative Price	Performance	Availability	Technology
Bosch	seat 150°C membrane ~ 350°C	low	medium	currently not available	Si
Denso	seat 150°C membrane ~ 400°C	low	medium	available in Japan	Si
Kulite	400°C	high	good	available*	SOI
Daimler-Chrysler Kistler	400°C	low	medium	prototypes available	SOI, SiCOI

* Currently available but not for automotive applications

Table 2.3 : Media Temperatures of Piezoresistive Sensors

Chip Technology	Chip Direct Exposure	Steel Membrane	Steel Membrane with Transmission Element
Si	150°C	200°C - 250°C	450°C - 500°C
SOI	350°C	400°C - 450°C	650°C - 700°C
SiCOI	500°C	550°C - 600°C	800°C - 850°C

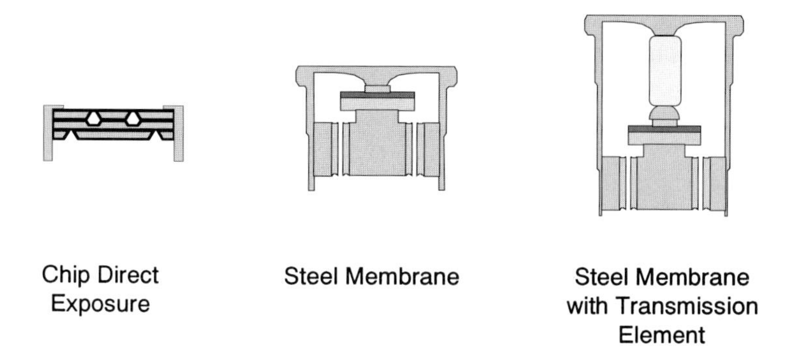

Fig.2.1: Schematic Diagram of Chip Technologies

3 Technical Approach

A forecast was made of new electronic features in cars by Delco Electronics [5]. Their predictions included the availability of combustion (or cylinder) pressure sensing in 2000, Model Based Control and individual cylinder control by 2005 and self-calibrating (adaptive control) engines by 2010. The Cylinder Pressure based EMS technology proposed in this paper will accelerate the introduction of these technologies into the market.
The approach taken in the AENEAS Project is to realise an EMS which fully utilises the potential of cylinder pressure sensing with intelligent control algorithms in order to obtain substantial improvements in cost, efficiency, comfort, emissions and reliability compared to conventional systems.

3.1 Sensor Technology

The AENEAS Project involves the application of innovative sensing technology. This is based on advanced R&D work carried out by DaimlerChrysler and Kistler on the development of inexpensive pressure sensing devices. The key to the technical breakthrough is the application of SiCOI and SOI technologies for such devices. These material systems have been shown to exhibit high performance and long term stability under harsh combustion environments thus providing a cost effective solution to the development of a CPEMS. This has not been possible with previous technologies.
Preliminary tests of SiC piezoresistors in which the SiC was epitaxially grown onto SOI substrates have shown good performance [6, 7, 8]. These have been shown to withstand high temperatures and provide accurate measurements over each combustion cycle.

Figures 3.1 shows the SiCOI pressure sensor and Figure 3.2 indicates the position of the pressure sensor in the engine.

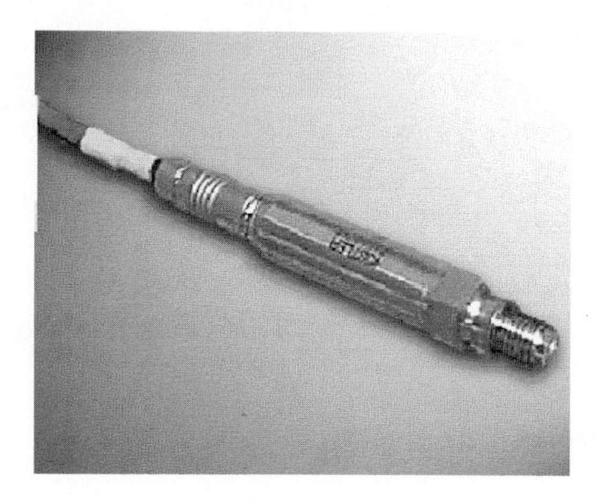

Fig 3.1: SiCOI Cylinder Pressure Sensor

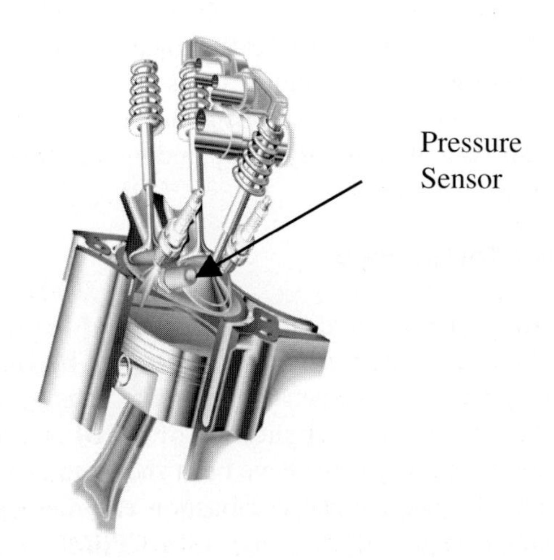

Figure 3.2: Position of Pressure Sensor in Cylinder

System for Radio-Controlled Car Clocks

Roland Polonio, TEMIC Semiconducters, Theresienstr. 4, 74025 Heilbronn
Reiner Häcker, HKW Elektronik, Industriestr 12 , 99846 Seebach/Thüringen

Outline

1. Introduction
2. On-board Car System
3. Time Signal Receiver
4. Clock
5. Summary

Keywords: Radio Controlled Clocks

1. Introduction

Today, more than ever, it is necessary that clocks (or more specifically, both systems that require the date and time or those that merely display the same), as well as displaying the exact date and time, should be as simple to operate as possible. Manual adjustments are time-consuming and are often not carried out. This is where an automatic system to change the time is required. In addition, the system should be able to automatically handle the twice-yearly change from Summer to Winter time, and back again, unnoticed by the user.

The radio-controlled clock is the answer to this problem – a system that receives time information from the official atomic clock, decodes it and thus can always be relied upon to supply the correct time and date.

At the time of going to press there are four official time transmitters, and a further one is planned for China.
DCF-77 transmits the official German time and covers a radius of approximately 2,000 km round Frankfurt/Main. MSF, stationed in Tettington near London, transmits the official GMT and covers a radius of approximately 1,500 km. WWVB is located at Fort Collins in Colorado and covers the whole of North America. It is responsible for UTC. One further transmitter, code name JG2AS, is situated in Japan and provides that country with the correct time.

All of these transmitters operate using long-wave between 40 and 80 kHz. Due to the very low frequency range, the magnetic waves spread out more or less accordingly to the Earth's surface and are able to penetrate buildings (provided the steel content is not too high). The expenditure of providing a time signal receiver has in the past few years been reduced to such an extent that a receiver can now be built into a normal wristwatch.

2. On-board Car System

The current, extremely complicated on-board electronics system of an automobile produces a nouse floor of interference in exactly that frequency range that is important for a time signal receiver. The processors and control elements located in the dashboard produce electrical interference, that could easily spread to an antenna that is built into the dashboard unit and hence only a limited reception would be possible. The integration of so many comfort features means that this interference does not cease with the removal of the ignition key. The processors continue for some time afterwards. Another source of interference that should not be overlooked are the higher currents flowing through connections in the dashboard. These currents produce magnetic fields, which can again cause interference to the antenna and a receiver. There is nearly always a fog of interference present from this mass potential that considerably reduces the reception. The above considerations lead one to the conclusion that the radio-controlled clock receiver must be developed differently from case to case.

3. Time Signal Receiver

The time signal receiver and analyzer is constructed as a separate module with an interface for data transfer and the power supply. This enables the module to be positioned where there is low interference and the highest possible fieldstrength in the automobile. This module can be fitted either in or even on the dashboard e.g. piggyback style where the interference allows. Should the quality of reception when fitted in the dashboard be unsatisfactory, then the module can be fitted in an isolated location further away, e.g. in the outside mirror. In all cases it must be ensured that reception of the magnetic components of the radio waves remains unaffected by the interfering components of the automobile electronics. Use of existing antenna, e.g. car radio aerials, has not been found to be effective due to insufficient quality of reception.

The module consists of a receiver, U4223B from TEMIC Semiconductors, a ferrite antenna and the microcontroller M44C092 from TEMIC Semiconducters, which is responsible for the signal analysis and the reception management as well as the data transfer of the current time and date to the on-board clock or timing system. The microcontroller is so designed that it produces no interference itself that could reduce the quality of reception.

The ferrite antenna can very easily be fitted according to the available space, thus building a compact unit. Even fairly small antennas can be produced with the necessary quality and thereby the required reception.

The receiver (TEMIC Semiconductors U4223B) is composed of a tuned radio-frequency receiver which gives the amplitude modulation of the time signal transmitter as an analog signal to an integrated A/D-converter. This latter converts the signal into a 4-bit digital signal. By means of special algorithms and through correlation analysis sufficient data can be extracted even from interfered or weak signals. Additionally, variation of the sampling rate of the A/D-converter can be achieved using the TEMIC circuitry. This enables further possibilities for effective analysis by employing the software developed by HKW-Elektronik.

The 4-bit microcontroller (TEMIC Semiconductors M44C092) has been specially designed to meet the requirements of the receiver (TEMIC Semiconductors U4223B). It is responsible for reception management (developed by HKW-Elektronik) that functions asynchronous to the on-board electronics. That means, when the on-board electronics produce no interference then the receiver for the time signal transmitter must be started. While the on-board electronics are active again, the module must make the information from the time signal transmitter available when called upon to the on-board electronics. Only an appropriate data interface must be designated. The appropriate protocol can be selected, taking into consideration the low current consumption of the timing processor, as the relevant interface on the clock module is easily programmable for different automobile types.
Due to the extremely low current of the controller, electromagnetic interference from the module to the receiver antenna is negligible. The total power consumption of the complete module is less than 50 μA, which a power supply buffer of several hours can supply with little effort. In this way the electronic function can be maintained for several hours even if the battery has been disconnected.

It is possible to produce variations of the module for use with the different transmitters (Japan USA, Europe), as these transmitters can be differentiated through their different modulation, protocol and frequency. As the interface itself already fulfills the requirements of the on-board electronics, there is no need to take these differences into account from the point of view of the on-board electronics. The same form of protocol can be used for all.

These measures ensure precise radio-verified date and time information is always available to the electronics. Consequently other date-dependent operations can be controlled (maintenance interval, error messages). As manipulation of this data via the protocol is extremely difficult, the system also offers security against manipulation of guarantee-relevant data.

Even where a system is already present, the interfaces can be arranged so that the radio timer module can always be fitted at a later date.

4. Clock

It goes without saying that every on-board clock can be fitted with a connection to the radio clock module where there is no need for this information by the on-board system itself. This converts each clock to a radio-controlled timer. As a variation to the digital clock, the possibility exists of fitting analog clocks with a connection. Another special feature available is a step motor module (HKW-Elektronik) with forwards and backwards movement of the hands.

5. Summary

Through a modular construction of the time signal receiver as individual unit, an asynchronous on-board electronic system can be achieved which provides the necessary undisturbed reception. The low power consumption enables power buffering over a longer period of time. The placement of the module is flexible and it can be located in any suitable position in the automobile. The synergy effect achieved between HKW-Elektronik and TEMIC-Semiconductors has made it possible to create a complete module with both the necessary software and hardware – specifically for the automobile sector.

Fuel Injection Engine Diagnosis

Patrick RIPOLL* **, Denis CAILLET *
Eric BENOIT**, Laurent FOULLOY**
* enterprise SEEM ZAC St-Estève lot No1 06 St Jeannet tel (33)04.92.12.04.80
seem@nicematin.fr
** Laboratoire d'Automatique et de Micro-Informatique Industrielle
LAMII/CESALP, Université de Savoie tel (33)04.50.66.60.40
{ripoll, benoit, foulloy}@esia.univ-savoie.fr
41 Avenue de la Plaine, BP 806, 74016 Annecy cedex FRANCE

Keywords: Engine Diagnosis, Control Strategy

1 Introduction

The strictness of antipollution constraints, the necessity of fuel economy or the development of comfort imply new control strategies for the automotive engine. With the increasing complexity of modern systems, diagnosis becomes a major theme for improving industrial process safety and reliability. This paper describes how the fuzzy logic can be used in diagnostic problems dedicated to the fuel injection automotive engine. The design of the diagnostic scheme is the following. Pertinent symptoms that respond to a given fault are generated. The residual computation detect whether a fault has occurs or not. Then, the analysis of these residuals with fuzzy logic operators perform the localisation procedure to determine the cause of the fault. Finally, the identification procedure is performed to determine the size of the fault. The decision is processed by fuzzy rules. In co-operation with Bosch company, over 2500 sensor data have been analysed for the purpose of diagnosing throttle sensor bias and manifold pressure bias. Results proved that the diagnostic tool is able to identify throttle sensor bias with a precision of 4 degrees and 130mbar for the manifold pressure. Then, an aided-computer diagnostic tool will be developed and future work will lead to the implementation of the method in an onboard diagnosis for autoadaptative injection control strategies.

2 Automotive Diagnosis

Due to the increasingly number of sensors in the automotive engine, onboard diagnostic tools (OBD) are one of the most important research field in automotive engineering. Detecting sensor faults remains necessary to keep the vehicle behaviour under nominal state. The most straightforward method is the threshold evaluation. Sensor monitoring algorithms mostly compare the sensor signal *e(k)* with predefined fault thresholds T_{iinf} and T_{isup} such as :

if $e(k) < T_{iinf}$ *or* $e(k) > T_{isup}$ then a fault is detected.

Such algorithms are not efficient since a small sensor fault affects the system, i.e. the inaccurate sensor value still belongs to the nominal range. Furthermore, effects of unknown inputs or disturbances are not taken into account with such thresholds. To avoid such non detection, we propose a new diagnostic algorithm based on fuzzy logic to perform small sensor fault detection. This report summarises the results of a collaboration between the Sud Est Electro-Mécanique company (SEEM) and the Laboratoire d'Automatique et de Micro-Informatique Industrielle (LAMII). At the present time, the application responds to only two sensor bias of the automotive engine : manifold pressure sensor and throttle angle position sensor. But the generic characteristic of the method will allow us to reuse it with different kinds of faults (sensor drifts, actuator faults...). These two faults have been chosen for their implementation facilities on the automotive engine. Their simulation have been provided by disconnecting the throttle angle sensor and sending a new throttle angle value to the injection calculator. For the manifold pressure, as the sensor is integrated in the calculator, a barometric bulb is assumed to replace the wrong manifold pressure. Test have been performed with a 2 litre injection engine (Peugeot 806) with a MP3.2 Bosch calculator.

3 Diagnostic Scheme

Diagnostic algorithms are fully represented in the literature (Willsky 1976) (Chow and Willsky 1984) (Frank 1990) (Patton and Chen 1991). The diagnostic approach considered consists of four steps as shown in figure1. For more details, see (Ripoll and al. 1998).

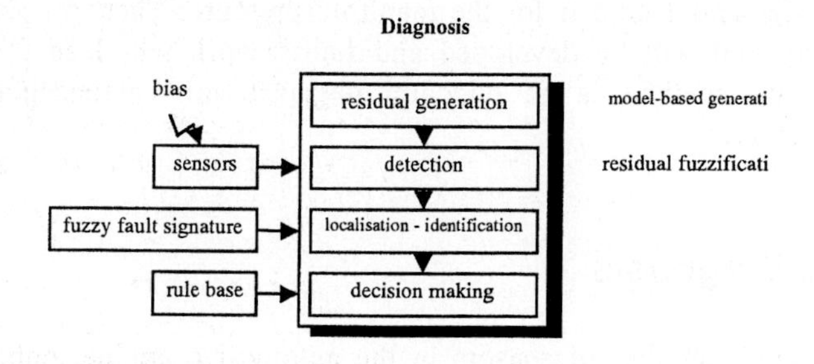

Fig. 1

3.1 Residual Generation

The first step of the method is the residual generation. Many methods are available: parity space or observers for example. Residuals correspond to fault indicators that respond to a fixed direction for a given fault and are null if no fault occurs. The residual set generated contains four equations. For this purpose, the mean compartmentalised model (Weeks and Moskwa 1995) was considered. Since desired fault detection concerned the throttle angle and the manifold pressure, the only subsystem of the air supply have been carried out. Structured set of residuals must be sensitive to some faults and insensitive to perform the isolation. Figure1, figure2 and figuer3 show respectively a 9 seconds residual time history for a non faulty system, a small throttle angle sensor bias (3°) and a small manifold pressure sensor (0.13bar).

3.2 Detection

The second step of the diagnosis consists of the detection to determine whether a fault has occurs or not in the system. The detection is a decisive step in the diagnosis and is equivalent to a threshold evaluation. Limit between non detection and false alarm can be improved with an adaptive fuzzy thresholds reasoning (Frank and Kiupel 1996). Another technique involves a residual fuzzification so that the decision making is done at the last step of the diagnosis. Membership functions can be computed by propagating the fault through the residuals (Ripoll and al. 1998).

3.3 Localisation

The problem in this section is to find which sensor is faulty. This is done by estimating each sensor bias. The isolation power is strongly dependant of the structured set of residuals. In order to enhance the structural properties of the incidence matrix, notion of fuzzy fault signatures are introduced by considering the sensitivity of each residual toward each fault. Then, a specific algorithm has been developed to provide an accurate comparison between fault signatures and residual vector. The identification of the fault (magnitude of the bias and time of the appearance of the fault) is performed by means of two specific numbers: a pessimistic value and an optimistic value of each bias sensor. These two weights allow the identification procedure to take into account the disturbances of the system or the uncertain sensor measures. Indeed, the model of the air supply has been established for ambient standard pressure and temperature so that for different ambient conditions, it needs a pressure and temperature correction. We assumed that these two measures are unknown, i.e. the there are no such onboard sensors. Nevertheless, an easily estimation can be done but this will introduce an uncertainty in the measures.

3.4 Decision Making

The last operation in the diagnostic scheme is the decision making. Fuzzy rules given by expert knowledge on the system decide whether a fault has occurs and return the magnitude and the time of appearance of the fault. Actually, the database contains four rules of type:

IF *$fault_k$_min is A_i* ***AND*** *$fault_k$_max is B_i* ***THEN*** *f_k is C_i*

where *$fault_k$_min* is the pessimistic measure of the kth fault, *$fault_k$_min* is the optimistic measure of the kth fault, f_k is the magnitude of the kth fault, A_i, B_i and C_i are linguistic terms. Such rules refine the initial fault range *[$fault_k$_min , $fault_k$_max]* calculated in 3.3. Figure5 and figure6 show the final results of the diagnosis for respectively a small bias in the throttle position and a small bias in the manifold pressure. The second case introduces a false alarm on the manifold temperature due to the poor isolating detectability matrix property. This false alarm has been also observed in another test using the hamming distance.

4 Conclusion

In this paper, a specific diagnostic algorithm has been developed to detect, isolate and identify sensor bias on the air supply system of the automotive engine. The use of fuzzy logic in the diagnosis gives very encouraging results. Small sensor bias on the throttle sensor is well identified whereas the manifold pressure induces a fault in the manifold temperature. Indeed, both sensor have the same fault signatures. This can be avoided by determining new residuals which are sensitive to one fault and insensitive to the other one. Perspectives will lead to the elaboration of a new engine model including fuel filling dynamics.

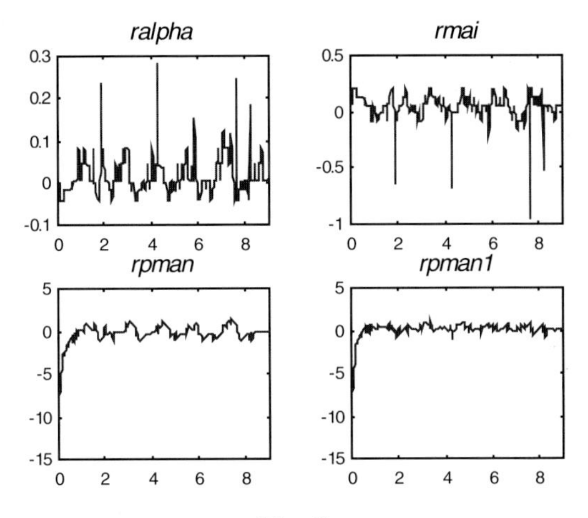

Fig. 2

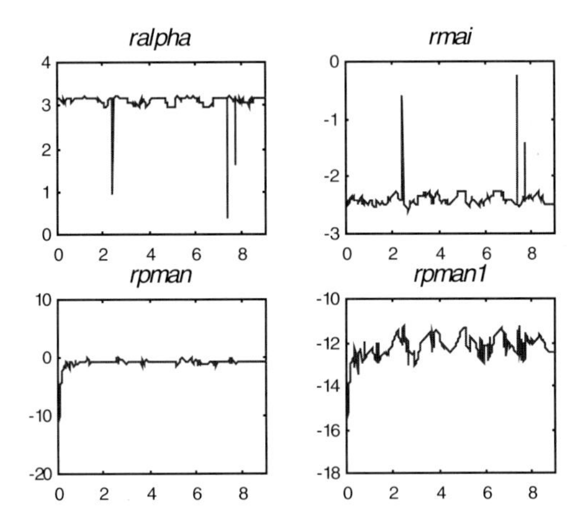

Fig. 3

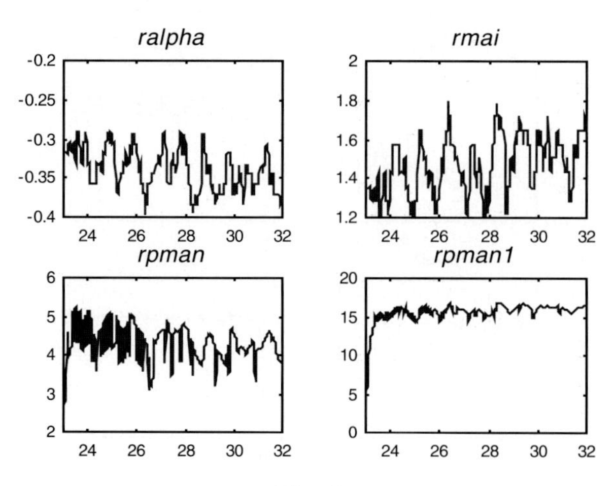
ralpha
rmai
rpman
rpman1
-0.2
-0.25
-0.3
-0.35
-0.4
2
1.8
1.6
1.4
1.2
6
5
4
3
2
20
15
10
5
0
24
26
28
30
32

Fig. 4

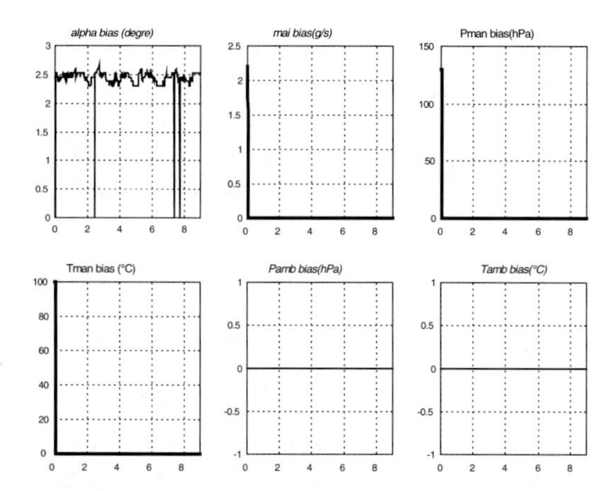
alpha bias (degre)
mai bias(g/s)
Pman bias(hPa)
Tman bias (°C)
Pamb bias(hPa)
Tamb bias(°C)

Fig. 5

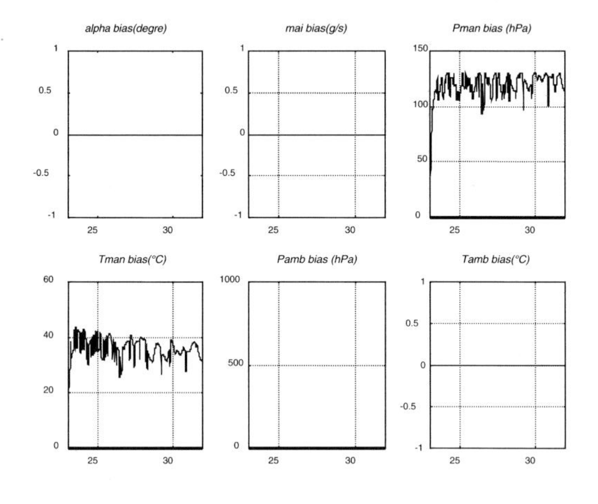

Fig. 6

5 Acknowledgement

We are grateful to the Bosch company for their contribution in this work and especially by placing a vehicle at our disposal and by helping us making the tests.

6 References

Chow EY, Willsky AS (1984) Analytical redundancy and the design of a robust failure detection system. IN: IEEE trans., Automatic Control 29 no 7.

Frank PM (1990) Fault diagnosis in dynamic systems using analytical and knowledge-based redundancy - A survey. In: Automatica, vol. 26, no 3, pp 459-474.

Patton RJ, Chen J (1991) A review of parity space approaches to fault diagnosis. In: IFAC Symposium SAFEPROCESS, Baden-Baden.

Ripoll P, Caillet D, Benoit E, Foulloy L (1998) Fuzzy implications handling fault detection and isolation problems. In: EUFIT, Aachen, Germany, pp.

Willsky AS (1976) A survey of design methods for failure detection systems. In: Automatica, vol. 12, pp 601-611.

Weeks R, Moskwa JJ (1995) Transient air flow rate estimation in a natural gas engine using a non linear observer", SAE 940759.

Monolithic Pressure Sensor System with Digital Signal Processing

J. P. Schuster, W. Czarnocki, X. Ding
Motorola Automotive and Industrial Electronics Group
Northbrook IL USA
B. Roeckner
Motorola Corporate Communications Research Lab
Schaumburg IL USA

Abstract

Most high volume applications of automotive pressure sensors have historically used some form of analog signal conditioning circuit to calibrate and compensate the silicon piezoresistive sensor element that has dominated many of these applications. This signal conditioning approach has often been implemented with some type of laser trimming to provide the appropriate circuit adjustments. Improvements on this approach have been introduced where electrical programming (e.g. fused links or nonvolatile memory) has been used to calibrate the circuit. However, even digital implementations of these calibration methods still rely on a core analog signal processor, thus providing only a discrete-analog solution to the sensor signal conditioning problem. This paper describes a monolithic pressure sensor integrated circuit that uses a custom, dedicated digital signal processor and nonvolatile memory to calibrate and temperature compensate a family of automotive pressure sensor modules for a wide range of applications. A specially developed digital communications interface permits calibration of a sensor module using the existing module connector pins after the module has been fully assembled and encapsulated. This approach eliminates any post-trim processing that could affect the sensor calibration. Module customization and calibration can be performed as an integral part of the end-of-line testing that is done at the completion of the sensor module manufacturing flow. Sub-micron CMOS circuit fabrication, bulk silicon micromachining, and wafer level bonding technology are uniquely combined to produce a cost-effective part which meets relevant automotive pressure sensor performance and durability criteria. Both digital and analog sensor outputs are available. This fully digital signal processing circuit provides programmable customer features and a significant degree of application flexibility without the system overhead and cost associated with a microprocessor implementation. It also provides a signal conditioning platform that can be applied to a variety of sensing technologies beyond just silicon piezoresistive pressure sensors.

Introduction

A key factor in the successful application of micromachined sensors to automotive applications is the ability to provide a truly interchangeable, ready-to-use sensor module to vehicle manufacturers and system suppliers. In order to do this, it is essential to provide a cost effective, robust signal conditioning circuit and manufacturable calibration method for converting the low level signal of the microsensor into a signal that is usable at the system level under a wide variety of harsh environmental conditions. The silicon micromachined pressure sensor is the most mature example of a microsensor used in automotive applications, and has evolved through several generations of signal conditioning designs and calibration strategies by various pressure sensor manufacturers.
Almost all pressure sensor signal conditioning methods that have been introduced into mass production have used some form of analog circuit. The typical calibration method for such products involves adjustment of the circuit by means of trimming either thick film or thin film resistors. Recently, variations on this analog circuit have been introduced using so-called “digital” trimming in which discrete values of the analog circuit parameters are chosen in the calibration process. However, this paper introduces a truly digital approach in both the signal conditioning and calibration methods used to process the microsensor output signal. This is done without the need for a microcontroller and the associated overhead and cost that would accompany such an approach.

System Architecture

A circuit design and trimming method have been developed which use a custom digital signal processing block that provides maximum desired functionality and accuracy within the silicon area and cost constraints demanded by automotive stand-alone sensor module applications. Similar to analog laser trimmed circuits, the digital programmable circuit also provides for span and offset calibration, as well as temperature compensation of span and offset. However, the digital circuit provides more capability to account for higher order inaccuracies in the sensor output, such as pressure non-linearity and non-linearity of temperature coefficients.
A block diagram of the programmable digital signal conditioning architecture is shown in Figure 1.

Specifically, an algorithm that mathematically represents the sensor output as a function of temperature and pressure is used to accomplish all of the calibration and temperature compensation tasks. This algorithm is implemented in the form of a custom DSP (Digital Signal Processor) circuit block that expresses the output of the sensor as the polynomial:

(Eq. 1) $V_o = a_0 + a_1V_t + a_2V_t^2 + (a_3 + a_4V_t + a_5V_t^2)V_p + (a_6 + a_7V_t + a_8V_t^2)V_p^2$

where, Vo is sensor module output,
a_0 through a_8 are constant terms,
V_t is the digital representation of the temperature signal
V_p is the digital representation of the pressure signal.

The values of coefficients a_0 to a_8 in Equation 1 are determined during the calibration and temperature compensation process. After the coefficients' values are determined, they are written into the sensor's non-volatile memory. A specially developed digital communications interface permits calibration of the sensor module using only the three existing connector pins which are integral to the sensor housing. This calibration process occurs after the module has been fully assembled and encapsulated, thus allowing the finished module to be both trimmed and tested at end of line. This eliminates the need to trim a partially assembled sensor module in the middle of the manufacturing process, which subsequently must undergo end of line testing to verify the calibration.

Fabrication

A high volume, mixed-signal, sub-micron CMOS process is used for the integrated circuit, and the sensor piezoresistor and cavity etch area are formed with process steps that are compatible with this existing process. Bulk micromachining, wafer bonding and dicing are performed and the individual monolithic sensor die are then assembled into completed pressure sensors at the electronic module assembly factory where they are customized for individual customer applications. This manufacturing strategy takes advantage of economies of scale and process learning curve effects, while maintaining sufficient flexibility to meet the needs of specific product applications.

Design Features

Designed for nominal 5V operation, this ratiometric programmable CMOS monolithic pressure sensor can withstand a DC overvoltage of 16 V, as well as survive ESD and other electrical transients. It has low current consumption, and is ratiometric to the supply voltage. Its output stage delivers a rail-to-rail output voltage swing with a load resistance of 10k, and can accept capacitive loads on the order of 0.033 mF. In order to meet certain electromagnetic compatibility (EMC) requirements, special means are provided within the integrated circuit to maintain satisfactory sensor operation at electromagnetic fields strengths that are typically encountered in automotive applications.

Other beneficial features include the ability to re-trim parts, and the capability to trim many parts essentially in parallel versus the sequential flow required by a laser trimmed analog part. Also, on-chip memory can provide information for part tracking through the manufacturing process to facilitate yield improvement activity. This part identification also provides useful part history in the event of a customer field return.

The custom DSP-based sensor system architecture provides the capability to provide customer-specific features such as programmable diagnostic clip levels and adjustable response time, in addition to unique transfer functions and accuracies for specific applications. The "digital ready" format of the sensor can provide an SPI output in addition to an analog voltage output.

Performance

In order to provide a truly interchangeable sensor module to the vehicle manufacturer, individual calibration and temperature compensation of each sensor is performed using a digital, programmable, trim algorithm. The output characteristics of a typical sensor are shown in Figure 2 for a manifold absolute pressure application. The digital signal processing method can provide a variety of transfer functions, as required by specific customer applications. Figure 3 shows the capability of the part to achieve +/-0.5% accuracy across a broad range of operating temperatures, with a maximum deviation of +/-1.0% at the -40 C and +125 C temperature extremes.

In order to maintain useful operation in a wide range of vehicle applications, sensor modules must be resistant to a wide variety of electromagnetic interference conditions. When subjected to a 200 Volt/meter field strength on a stripline, output deviations of this sensor module are typically less than 30 millivolts. The sensor can also survive direct RF injection levels of up to 2.0 watts, where it still maintains its functionality.

Different automotive pressure sensor applications sometimes require different output response times, depending on the control system and system algorithm that uses the sensor's output. In order to accommodate this application-specific requirement, the monolithic sensor's signal processing block contains a digital filter which permits programmable selection of an application-specific response time.

Applications

The flexibility of the programmable monolithic pressure sensor chip design (Figure 4) allows it to be used in a wide variety of automotive applications: manifold absolute pressure, combined inlet air temperature and manifold absolute pressure, barometric pressure, evaporative fuel tank emissions, differential exhaust gas pressure, fuel injection pressure, and engine oil pressure.
Unlike capacitive, surface-micromachined CMOS integrated pressure sensor designs, the bulk micromachined piezoresistive design can easily be used for gauge or differential pressure measurement applications, which constitute a major share of the automotive market applications. Consequently, this monolithic sensor element was designed to meet a variety of pressure ranges and measurement formats with just one integrated circuit, only requiring minor adjustments to the sensor cavity mask to adjust for diaphragm size and thickness as different product applications require. Figure 5 shows examples of products using the monolithic sensor IC in absolute, gauge, and differential implementation.

Conclusion

A CMOS programmable sensor signal conditioning system has been developed which overcomes the limitations of existing analog designs and calibration methods. This cost effective, manufacturable design uses digital signal processing to simplify the sensor module production and calibration process, and is flexible enough to meet the performance targets for a wide variety of automotive pressure sensing applications. Additionally, this integrated circuit provides a core technology platform for a family of mixed-signal sensing systems which can calibrate and temperature compensate a variety of sensor element output signals beyond just pressure sensors.

Acknowledgments

The cooperative development of this product was made possible by an extensive team of people throughout Motorola, from a variety of technical disciplines and organizations: Corporate Communications Research, Automotive and Industrial Electronics Group / European Automotive and Sensors Division (sensor element manufacturing, sensor module manufacturing, and sensor module development teams), Semiconductor Products Sector (MOS-8 and Bipolar 6 wafer fabs, and Powertrain Products Operations team), and Motorola Manufacturing Systems. The contribution of the following individuals was particularly instrumental in determining the initial technical feasibility of this device: Rick Brownson, Dave Scott, Tim Rueger, Brendan Doorhy, and Carolyn Johnson.

List of Figures

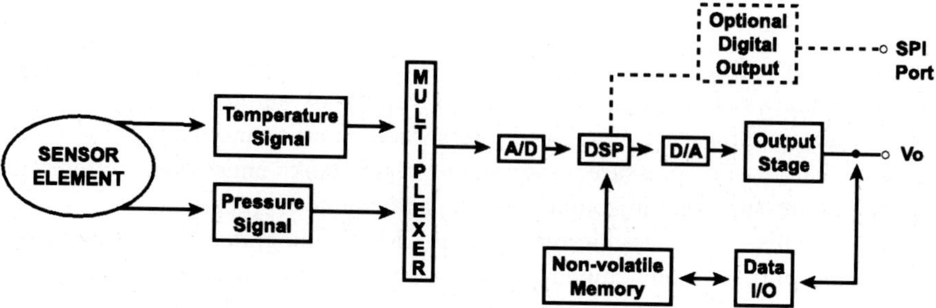

Fig. 1: Programmable sensor: digital signal processing architecture.

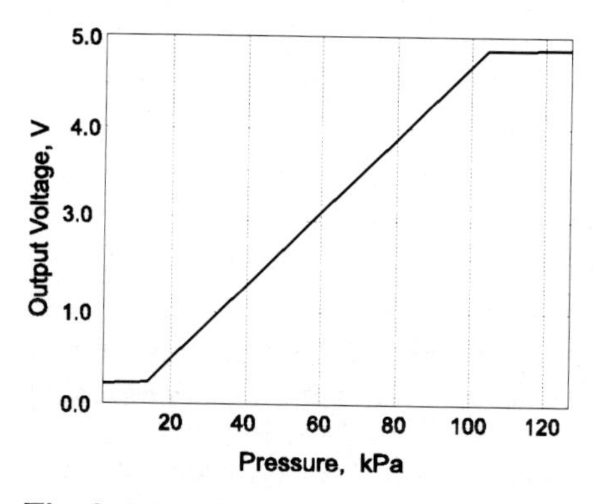

Fig. 2: Monolithic pressure sensor transfer function.

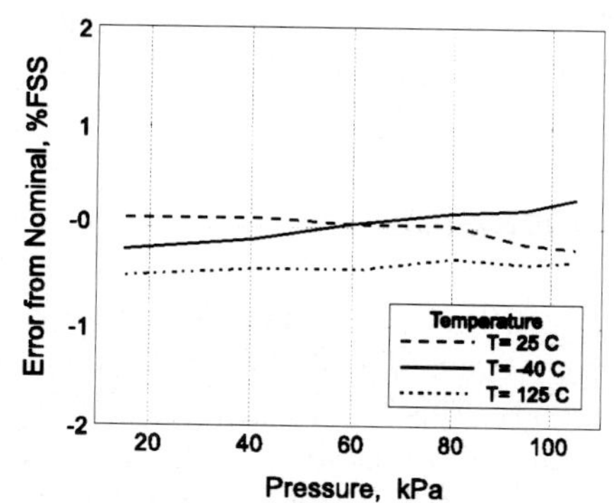

Fig. 3: Monolithic pressure sensor accuracy.

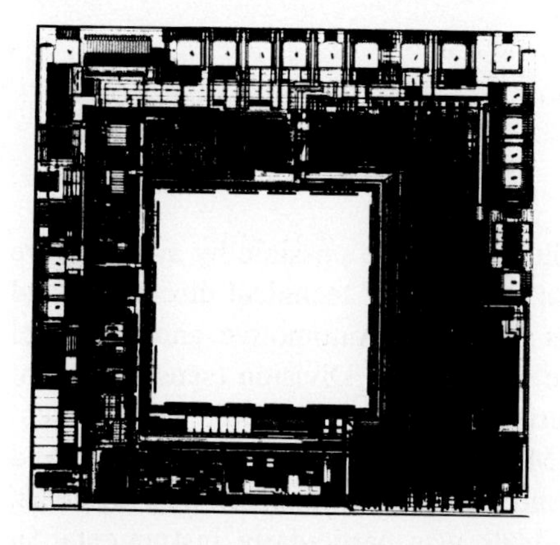

Fig. 4.: Monolithic pressure sensor integrated circuit.

Fig. 5: Manifold absolute pressure, exhaust gas pressure, and oil pressure applications.

3.2 Intelligent Control and Diagnostics

CPEMS relies on close monitoring of the engine's performance together with Model Based control of individual cylinders. This results in a number of improved engine functions. For example, since the cylinder pressure sensors enable direct monitoring of the engine's combustion, knock detection and control is substantially improved. Also Model Based control algorithms will be applied, enabling the system to adapt and therefore compensate for in-service wear over the life of the engine. Furthermore, this new technology will be transferable to new engine concepts such as Gasoline Direct Injection (GDI) and Variable Valve Lift Control for further improvements in performance.
The following is a list of some of the possibilities in control and diagnostics:

1. The air mass flow rate through the engine can be obtained by estimating the mass of air inducted into each of the cylinders. Any possible errors in the system, such as leakage or throttle body variability, will have no influence on the accuracy of the estimated air mass.
2. The timing of the ignition can be implemented as a feedback control loop for each cylinder. This provides greater flexibility in on-line adaption to ensure optimal timing during the life of the engine.
3. Pressure based knock detection will be more accurate than conventional systems thus leading to more effective knock control.
4. OBD will be improved since the combustion pressure gives a closer insight in to the burning process. Possible applications include peak pressure detection and control, compression monitoring and accurate misfire detection.
5. The position of cylinder one can easily be detected therefore replacing the cam shaft sensor.
6. To ensure that all cylinders produce the same crankshaft torque, a pressure based roughness control can be implemented.

CPEMS will replace a number of sensors used in conventional EMSs. As indicated by 1, 3 and 5, these will be air mass flow, knock and cam sensors. This will provide substantial cost benefits which will outweigh the additional cost of cylinder pressure sensors.

CPEMS will provide the following benefits over conventional systems :

- Better controlled combustion giving lower emissions, faster catalyst light-off and smoother running
- Improved sensing leading to a reduced number of sensors and hence cost
- Faster sensing
- Reduced number of engine maps due to the use of Model Based control
- Improved adaptive capabilities since the control algorithms will be Model Based
- Increased suitability for Model Based control hence reduced calibration effort and greater scope for novel control methods
- Improved OBD
- Reduction of engine material costs since engines will be built closer to their design limits

This approach will also provide the following benefits over ion current sensing systems:

- Pressure information valid for whole cylinder
- Greater scope for applying Model Based control since more physically based models can be realised
- Direct measure of cylinder pressure
- Easier to locate peak pressure
- Faster processing of information since reliable pressure measurements can be obtained in one engine cycle
- The use of EGR and different fuel additives is made possible since the sensing methodology is independent of chemical properties
- Sensed information can be obtained over whole engine and ambient operation (including full engine cycles).
- Transferable to diesel and natural gas engines
- More easily transferable to new engine concepts, such as Gasoline Direct Injection (GDI) or Variable Valve Lift Control

The following list contains the main control and diagnostics aspects which would be covered in the project.

1. Misfire Detection
2. Knock Detection
3. Cylinder 1 Detection
4. Mass Airflow Estimation
5. Combustion Stability
6. Start Control

7. Sensor Error Handling
8. Electronics Error Handling

Mass airflow can be monitored by estimating the mass of inducted air during the pumping loop. Preliminary in-house studies using Adaptive Kalman Filtering [9] have yielded promising results. This algorithm, based on the maximum likelihood method in which a physically based model was applied, has demonstrated rapid and stable convergence of the estimated mass airflow to a measure signal.

As mentioned in Section 3.1, combustion stability can be improved by individual control of the combustion in each of the cylinders. Previous work has been carried considered the application of Neural Networks to the problem of optimal ignition control [10, 11]. Cylinder pressures were sensed and monitored by a Neural Network to determine the optimal ignition timing. Successful results were obtained on a Mercedes Benz E-Class vehicle using a linear controller. There are plans to use a recurrent Neural Network to control the ignition timing for improved transient response.
Figure 3.3 illustrates the control possibilities with CPEMS technology.

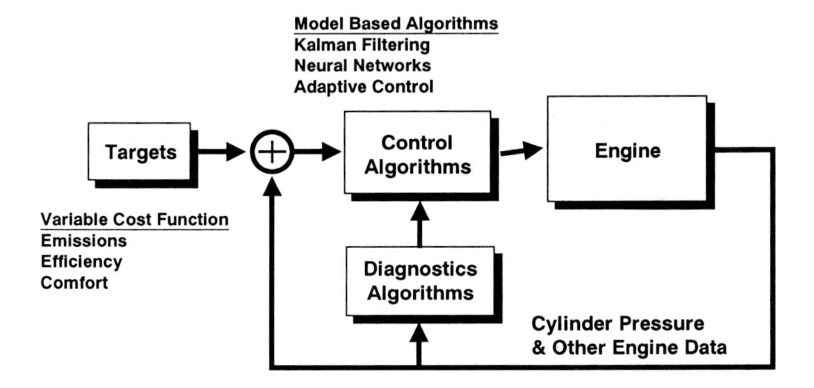

Fig. 3.3: Control Possibilities with CPEMS

4 Technical Development

The technology input to the overall project is in the following areas :

- Fixing specifications and automotive boundary conditions for CPEMS
- Adaption of existing cylinder pressure sensors for the required application
- Adaption of existing algorithms to system requirements
- Fulfilling increased quality and lifetime requirements of hardware components

For the overall system, engine, sensors and electronic component specifications will be finalised. A detailed plan will then be drawn up of the complete system including electronic architecture.

With regards to the sensors, high pressure sensor prototypes will be adapted according to system requirements. This includes scaling up sensors for low and high pressure ranges. The electronic amplification and compensation, and mechanical configuration will be optimised for installation into a cylinder head. A series of tests will then be carried out at Kistler in order to assess sensor stability and lifetime.
Concerning the supply of algorithms, control and error handling algorithms will be adapted for the specific vehicle application. Advanced simulation tools will be utilised for adapting and testing the algorithms prior to implementation. A rapid prototyping system will be used interfacing with a standard ECU.
The CPEMS technology will be assimilated at Ricardo under the technical leadership of the Vehicle Control and Electronics Department. The objectives of this work area are as follows :

- To realise the benefits of CPEMS over conventional EMS technology
- To establish automotive testing facilities close to serial production
- To quantify the financial and technological effort for integrating CPEMS into existing production lines
- To realise the compatibility with existing production methods and control architectures of engines
- To show possibilities of transferring the developed systems into other types of engines

The CPEMS technology will be assimilated on an engine testbed and a demonstrator vehicle. The application, to be supplied by DaimlerChrysler, will be a 3.2 litre V6 dual spark ignition engine to run in a Mercedes Benz vehicle.

5 Conclusions

This paper gives a brief overview of combustion monitoring techniques and the benefits of CPEMS technology. Previous work has shown the potential benefits of applying SiCOI and SOI technologies for sensing cylinder pressure and the potential of advanced control algorithms which have previously been tested on an engine. The AENEAS Project aims to combine these technologies to develop a cost effective solution to CPEMS and to realise the many benefits of this technology.

Acknowledgements

This work is financially supported by the European Commission DGXII/D as Innovation Programme IN301056I. The permission of DaimlerChrysler, Kistler and Ricardo to publish this paper is also acknowledged.

References

[1] Asano M, Kuma T, Kajitani M, Takeuchi M (1998). Development of New Ion Current Combustion Control System. SAE 980162

[2] Balles E, VanDyne E, Wahl A, Ratton K, Lai M (1998). In-Cylinder Air/Fuel Ratio Approximation Using Spark Gap Ionization Sensing. SAE 980166

[3] Gilkey J (1984). Fuel-Air Ratio Estimation From Cylinder Pressure in an Internal Combustion Engine. PhD Dissertation, Stanford University

[4] Wibberly P, Clark C (1989). An Investigation of Cylinder Pressure as Feedback for Control of Internal Combustion Engines. SAE 890396

[5] MIRA Engine Review (1997)

[6] Krötz G, Wondrak W, Eickhoff M, Lauer V, Obermeier E, Cavalloni G (1997) New High-Temperature Sensors for Innovative Engine Management. AMAA 97, Springer-Verlag Berlin Heidelberg New York, pp. 223-230

[7] Eickhoff M, Möller H, Krötz G, v. Berg J, Ziermann R (1998) A High Temperature Pressure Sensor Prepared by Selective Deposition of Cubic Silicon Carbide on SOI-Substrates. Paper presented at E-MRS Spring Meeting 1998 Strasbourg, France

[8] Ziermann R, v.Berg J, Obermeier E, Eickhoff M, Krötz G (1997) A High Temperature Pressure Sensor With Beta-SiC Piezoresistors on SOI Substrates. Technical Digest International Conference on Solid-State Sensors and Actuators, Chicago, USA, p.1411

[9] Hart M, Ziegler M, Loffeld O (1998) Adaptive Estimation of Cylinder Air Mass Using the Combustion Pressures. SAE 980791

[10] Müller R, Hemberger H, Baier K, Gern T (1996) Control and Diagnostics in Automotive Applications. 96A4047

[11] Müller R, Hemberger H (1998) Neural Adaptive Ignition Control. SAE 981057

An MCM Microcontroller for Automotive Applications

Arzu Simsek, Wilhard Strohschein, Herbert Reichl
Fraunhofer Institut FhG-IZM
Gustav-Meyer-Allee 25, 13355 Berlin, Germany
Tel: +49 30 464 03 144; Fax: +49 30 464 03 111
E-Mail: simsek@izm.fhg.de
http://www.izm.fhg.de/europractice/

Keywords: Multi Chip Module, Packaging

Abstract

A fast microcontroller in a high density Multichip Module (MCM) package is developed which is EMC compliant and suitable for automotive and industrial applications. The 16-Bit Microcontroller MCM was realised in laminate technology with wire-bond chip interconnection and a custom-specific package in reflow-soldered edge clip gullwings in standard QFP-outline. The EMC-constraints were easily solved by shielding the MCM with its own ground-plane. Reliability and cost-targets are met by the widely-structured, low-cost 4 layer laminate technology. It can be used in a wide range of industrial and automotive applications. A number of variants of the MCM design are now under development

Introduction

The vehicle control functions typically range from security systems to suspension, transmission, instrumentation and comfort control. Applications include e.g. electric seat contol, audio systems, cruise control, ignition systems, car alarms, power windows, keyless entry, air bags, etc. Consequently, over 60 onboard micropocessor-based control units MCU may be present on modern vehicles. These trends need application-tailored solutions such as fast microcontrollers in high density packages. In this paper a 16 Bit Microcontroller MCM (MCM16μC-A) is presented which is suitable for numerous automotive and industrial applications.

The electromagnetic compatibility EMC and the requirement of reliability and quality were the driving forces behind the development of a unit in the form of an

MCM. MCM technology offered the possibility of producing a module that met the electrical performance and reliability specifications.

Product Description

Integrating the Siemens SAB-C167CR-LM microcontroller with Flash, SRAM and EEPROM memory the MCM provides full-CAN / basic-CAN bus interface specification 2.0B (active). Figure 1 shows the block-diagram.

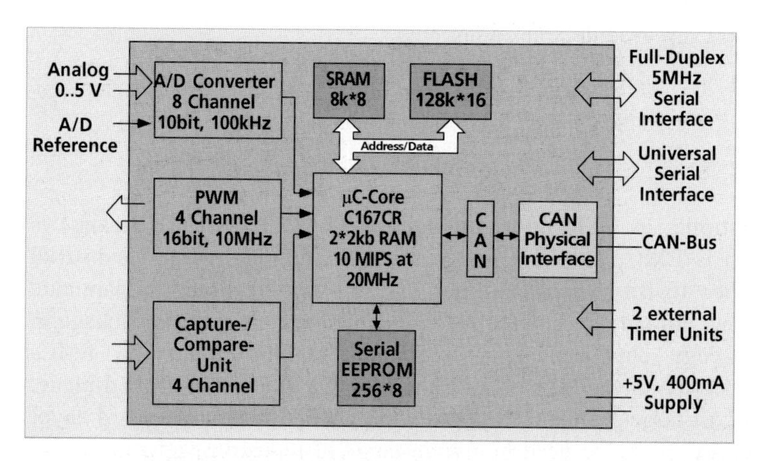

Fig. 1.: Block-diagram of the 16-Bit-Microcontroller MCM

The C-167CR-LM Microcontroller contains several integrated units like A/D converter, PWM and Capture/Compare units and CAN-Bus Controller (blue colour). Three additional bare dice provide program- and data-memory. An additional die is the CAN-Bus physical layer PCA82C250U. All dies are in the so called **K**nown **G**ood **D**ie quality or near KGD due to several wafer tests.

The type of passive components depends on substrates etc. Thermal cycling and thermal shock together with the automotive temperature range due to a position near the engine should not decrease the life cycle. Low-noise AD-conversion is necessary. Full internal power supply decoupling is preferred. The very strong EMC-constraints can only be reached by keeping the digital busses with high-frequency signals (e.g. data- and address-bus plus chip-select lines) inside the MCM. 13 resistors has to be integrated. Among other things they preselect a bootstrap loader which gets his data by using a serial interface.

Compared with a conventional system on a PCB, the 16 Bit µ-Controller MCM dramatically reduces size and weight (s. Fig.2).

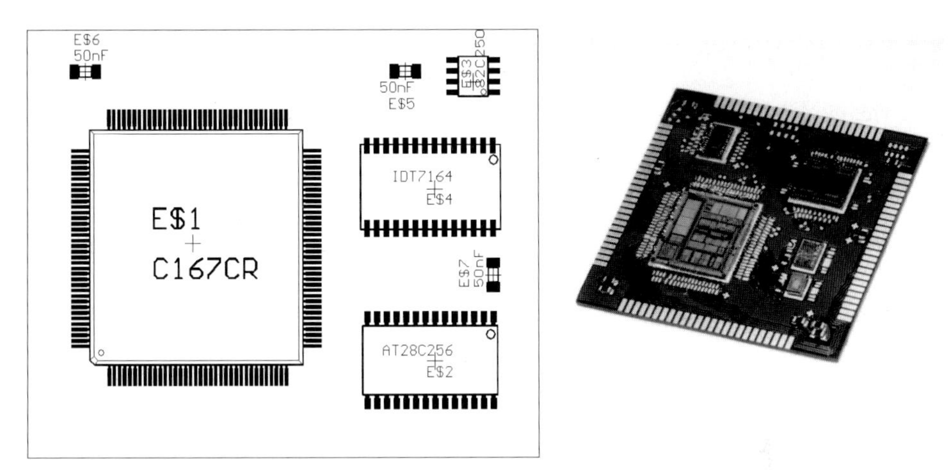

68 x 51mm² = 3468mm² 24 x 24mm² = 576mm²

Fig.2: Size comparison; MCM has about 6 times less area than the PCB

Materials and Processes Evaluated

There are 3 alternatives for the substrates:

- Advanced MCM-L FR-5 in different line and via dimensions with 2 signal layers and additional assembly- and ground-layer, size 24 x 24 mm². Edge-Clip Gullwings are used due to thermal shock problems of BGA packages to compensate the TCE mismatch resulting of special board TCE.
- Thin-film MCM-D on ceramic carrier with 2 layers. The size of the P-QFP package is 20 x 20 mm² with thin dies and without SMT capacitors. Dies of normal thickness and SMT-devices are only possible in 24 x 24 mm² package.
- Ceramic MCM-C: The size of 28 x 28 mm² and the cost were less favourable than the MCM-L substrates, and the application does not need to be in an hermetic package so the MCM-C option was not considered further.

By using a MCM-L substrate, the optimum combination of size-reduction, very good electrical properties and low weight can be achieved. The IC connections were made by wire-bonding. A thin-film module with similar assemby technology (aluminium wedge bonding) reduces the number of layers (MCM-L: 4 layers, MCM-D: 2 layers) but due to the assembly technology and problems by integrating SMT-capacitors the size reduction was low. A further size reduction is promised by using Flip Chip technology which has the potential to increase the electrical performance and to have a better reliability. In large volumes Flip Chip technology can be cheaper than the wire bonding.

Potential Applications

The module is designed flexible for general industrial control applications such as sensor and actuator interface providing decentralised processing capabilities in a highly integrated package. There are seven distinct markets by which the MCM16μC-A can be applied for significant cost/performance improvements. The seven areas are:

- Automotive applications
- Industrial control applications
- Sensor interfaces
- Controller for power switches
- Smart CAN bus components
- Equipment control
- High environmental requirements

Especially when using the MCM in high performance systems, the cost of the system decrease as there is no need for several high performance packages anymore.

Conclusion

In this paper, a 16-bit microcontroller MCM is presented which can be used in numerous applications where reliability, space and cost are severly restricted. A number of variants of the MCM design are now under development

References

[1] W. Strohschein, A. Simsek, H. Reichl: A 16-Bit Microcontroller MCM-L for Automotive and Industrial Applications compared with MCM-D Technology. IEEE Workshop on Microsystem Packaging Techniques and Manufacturing Technologies, 4-5 May 1998, Brugge

[2] F. Ansorge, K. F. Becker et.al: Assembly & reliability testing under harsh environment conditions for the qualification of MCM-L technology for automotive applications. ITAP98, Feb 1998 Sunnyvale

[3] A. Ostmann, G. Motulla et.al.: Low cost techniques for Flip Chip Soldering. Proc. SMI Conf. 96, San Jose

List of Contact Addresses

Matthias Aikele
TEMIC Sensorsysteme
Phone +49 89 607-25552
Fax +49 89 607-28545
email matthias.aikele@temic.de

Karl Billen
I.E.E. International Electronics & Engineering Luxembourg
Phone +352 42 4737-315
Fax +352 42 4737-200
email karl.billen@iee.lu

Jeff Casazza
c/o Andreas Berndorfer
Intel GmbH
Phone +49 89 99 143-638
Fax +49 89 99 143-479
email andreas.berndorfer@intel.com

Olivier Clair
Renault
Phone +33 1 34957-645
Fax +33 1 34957-722
email olivier.clair@renault.fr

Joël Duhr
Delphi Automotive Systems Luxembourg S. A.
Phone +352 5018-330
Fax +352 5018-780
email duhrj@pt.lu

Robert Frodl
Ruf-Electronics GmbH
Phone +49 8102 781-440
Fax +49 8102 781-411
email robert.frodl@ruf-electronics.com

Wolfgang Gessner
VDI VDE Technologiezentrum Informationstechnik GmbH
Phone +49 3328 435-173
Fax +49 3328 435-216
email gessner@vdivde-it.de

Roy Grelland
SensoNor asa
Phone +47 3303 5055
Fax +47 3303 5005
email roy.grelland@sensonor.no

R. E. Hardt
Industriekontor Rolf Hardt
Phone +49 211 71-2828
Fax +49 211 71-3186
email

Dan Haronian
Tel-Aviv University
Phone +972 3 640-6414
Fax +972 3 641-0189
email haronian@eng.tau.ac.il

Manfred Klein
DaimlerChrysler AG
Phone +49 731 505-2055
Fax +49 731 505-4103
email
m.klein@dbag.ulm.daimlerbenz.com

Laurent Lévin
Technocentre RENAULT
Phone +33 1 3495-7578
Fax +33 1 3495-7719
email laulev@club-internet.fr

Tarek Lulé
Silicon Vision GmbH
Phone +49 271 890-9660
Fax +49 8909675
email lule@siliconvision.de

Max Monti
CSEM SA
Phone +41 32 7205-572
Fax +41 32 7205-720
email max.monti@csemne.ch

Andy Noble
Ricardo Consulting Engineers Ltd.
Phone +44 1273 455611
Fax +44 1273 464124
email adnoble@rce.ricardo.com

Mikio Nozaki
Nissan Motor Co. Ltd.
Phone +81 45 505-8436
Fax
email m-nozaki@mail.nissan.co.jp

Marco Pizzi
Fiat Research Center
Phone +39 11 9023520
Fax +39 11 9023-673
email m.pizzi@crf.it

Roland Polonio
Temic Semiconductors
Phone +49 7131 67-2102
Fax +49 7131 67-3050
email roland.polonio@temic-semi.de

Frank Rehme
DaimlerChrysler Aerospace AG
Phone +49 731 392-4540
Fax +49 731 392-5465
email frank.rehme@vs.dasa.de

Detlef Ricken
VDI VDE Technologiezentrum
Informationstechnik GmbH
Phone +49 3328 435-242
Fax +49 3328 435-256
email ricken@vdivde-it.de

Patrick Ripoll
LAMII/CESALP
Université de Savoie
Phone +33 450 66 60 40
Fax +33 492 12 02 01
email ripol@esia.univ-savoie.fr

Wolfgang Robel
Reitter & Schefenacker GmbH & Co. KG
Phone +49 711 3154-168
Fax +49 711 3154-256
email ve1.rus@t-online.de

Ralf Schellin
Robert Bosch GmbH
Phone +49 7121 354172
Fax +49 7121 354173
email Ralf.Schellin@rt.bosch.de

John P. Schuster
Motorola
Phone +49 847 480 8153
Fax +49 847 205 3883
email gusr144@email.mot.com

Arzu Simsek
Fraunhofer IZM
Phone +49 30 46403-144
Fax +49 30 46403-111
email simsek@izm.fhg.de

Peter Sommerfeld
Philips GmbH
Phone +49 2151 576-311
Fax +49 2151 576-308
email Peter.Sommerfeld@kre.mimo.philips.com

Bob Sulouff
Analog Devices Inc.
Phone +1 617 761-7656
Fax +1 617 761-7607
email bob.sulouff@analog.com

Sebastian Toelg
EG&G Heimann Optoelectronics GmbH
Phone +49 611 492-271
Fax +49 611 492-228
email sebastian_toelg@egginc.com

Anthony Truscott
Ricardo Consulting Engineers Ltd.
Phone +44 1273 794057
Fax +44 1273 794563
email AJTruscott@rce.ricardo.com

Dung Tu
Texas Instruments
Phone +49 8161 80-4641
Fax +49 8161 80-4588
email dung-tu@ti.com

Sergey Y. Yurish
Institute of Computer Technologies
Phone +380 322 9716-74
Fax +380 322 9716-41
email syurish@mail.icmp.lviv.ua

List of Keywords

Keyword	Page
ABS	215
ACC	173
Accelerometers	225, 251
Active Microsensor	215
Adaptive Measurement	215
Angular Rate Sensor	239
Antennas	195
Autoadaptive Image Sensor	183
Automotive Radar	195
Automotive Vision System	183
Broadband Communication	195
Communication	13, 25
Control Strategy	289
CVT	97
Deep Reactive Ion Etching	225
Droplets	87
Drops On Demand	87
DSP	297
DSRC	63
Electrostatic Actuator	271
Engine Control	71, 97
Engine Diagnosis	289
Engine Management	71, 279

Entertainement 13, 25
Flow Sensors 199
Force Sensing Resistor 133
Fuel Economy 97
GSM - Based Services 13
GATS 13
Gyroscope 261
Handsfree Systems 43
High Dynamic Range Image Sensor 183
High Temperature Soldering 111
Hybrid Circuits 111
Image Sensor 183
In-Cylinder Measurement 279
Information 25
Initial Sensors 199
Injection Moulding 109
Intelligent-Rear-Light 147
Ionization Current 71
Laser Radar 173
Lateral Integrated Surface Micromachining 225
Linear Motion Rotative Motion 271
Mass Production 109
Media Compability 121
Microejection 87
Microphones 43
MIM/CIM 109
Monolithic Pressure Sensor 297
Multi Chip Module 309
Multi-Axial Sensor 251

Navigation	25, 261
Obstacle Detection	173
Occupant Classification	133
Packaging	111, 121, 195, 309
PC-Technology	25
PET	271
Powertrain	97
Pressure Sensor	121, 279, 297
Radio Controlled Clocks	285
Random-Modulation	173
Range Sensor	173
Resonant Sensor	251
Rollover Sensing	239
Rotative Motion	271
Scanners	173
SCREAM Process	199
Silicon Micromachining	251
Smart-Airbag-Systems	133
Surface Micromachining	239, 261
TFA Technology	183
Time-Of-Flight	173
Traffic Management	63
Transponder Technology	63
VDC	261
Voice Links	43
WAP	13